商务印刷品设计

COMMERCIAL PRINTS DESIGN

（塞尔维亚）伊利亚·德拉吉斯克
伊戈尔·米拉诺维奇/编
常文心/译

Annual Report
Booklet
Catalogue
Brochure
Promotional Leaflet
Annual Report
Booklet
Catalogue
Brochure
Promotional Leaflet

辽宁科学技术出版社

# CONTENTS 目录

| | |
|---|---|
| Preface<br>前言 | 004 |
| Annual Report<br>年度报告 | 008 |
| Booklet<br>广告册 | 055 |
| Catalogue<br>型录 | 120 |
| Brochure<br>商务宣传册 | 201 |
| Promotional Leaflet<br>商品促销单 | 257 |
| Index<br>索引 | 268 |

# PREFACE

Nowadays, design is as important as it was many centuries ago since man started creating and coming up with ingenious ideas in all spheres of life. Practical solutions that enhance and make life easier, had been instantly accepted by the community and still are of immense importance to human race. Man's urge for visual communication through drawings dates from the Stone age, when people decorated the stone walls of caves, which were therefore used as a medium for worshiping the culture-specific deities so as to have their prayers answered.

Little has been changed since then. Visual representations of people's inner thought processes are still the primary tool of communication on the whole planet, if not in the entire universe, because only by seeing, can we confirm its existence with certainty.

Given that we live in the age of consumerism, we are constantly being exposed to a myriad of information on a daily basis, just as the waves pound upon the seashore. It has been proven that nearly seventy percent of external information is in fact obtained through the eyesight. Hence, graphic design has the leading role in adequately transmitting those pieces of information that are intended for consumers. Consequently, we, as observers, reach for those products whose esthetic quality is at an elevated level. This choice is the result of certain mental processes in our mind, the existence of which we are usually unaware of; but there is the so-called subconscious thread that connects us with a certain product, something that "coerces" us into making the choice and going for it.

Products that are of sleeker design have a higher market share, are sold faster, and obtain larger profits. Before the very designing process, it is of the utmost importance to acquire a deeper knowledge and understanding of the group the product is aimed at, and adjust the design towards it. Graphic design has a key role in the marketing. Most people are usually exposed to concoctions of graphic design used for promotional purposes- posters, billboards, brochures, catalogues, leaflets. Apart from these forms of graphic design, people also encounter others, such as magazines, packaging, visual identities of companies, book covers...

One of the first steps in designing is the layout of elements on a format. Format is the space on which photographs, illustrations and the text should be placed in such a way as to enable the recipient of the message to easily perceive the information. The amount of information used on a specific format can vary, and may be small or vast. Tasks given to a graphic designer are various, and can either be rather complex and daunting or simple, the former requiring the craft and dexterity of a particular designer, or the whole creative team, to solve problems.

As complex as the amount of information may be, the result of a sound layout of elements should allow the perceiver to find out the desired piece of information without obstacles. Auspicious layout is applied in both printed and electronic media, basically, wherever graphic assistance is required.

Apart from a functional solution, the task of the designer is to pay attention to the level of aesthetics when dealing with the challenge. A fairly crucial item when arranging elements is the existence of unique visual identity of designed material, and recognition of certain patterns in page layout, that need to be unified across product palettes. This can be achieved by the means of a defined network of lines, which empowers precise and quick positioning of the elements on the format. On one hand, the network alleviates the positioning of the elements, whereas, on the other hand, it constrains and somewhat unifies the appearance, thus making dull composition of the final solution. Subtle touches, which are used when designing, are the reflection of experience and intelligence of the design team, and they make the difference between a mediocre and a world-shattering solution.

The very layout of the format can be solved instinctively, however, working in the field of graphic design

# 前言

如今,设计与几百年前人类开始在生活的各个领域运用创意理念时一样重要。能够使生活更加方便的务实解决方案很容易被人们所接受,同时在人类历史长河中发挥了极为重要的作用。人类运用图画充当视觉传达的手段可追溯到石器时代,在当时,人们用图画装饰洞穴内的石墙,并将它们作为图腾崇拜的一个媒介,希望神灵能够通过这一媒介,听到他们的祈祷。

至今,设计的地位并没有因为社会的发展而动摇。人类内心所想的视觉再现仍然是地球上视觉传达的基本工具,因为,只有真正所见,我们才能确信事物的存在。

现如今的时代是一个注重消费的时代,我们每天都在不断接触各种各样的信息,正如长江后浪推前浪。实践证明,有近百分之七十的外部信息通过视觉获得。因此,平面设计在以消费者为导向的信息的传递中扮演了重要的角色。因此,我们作为观察者,倾向于选择那些具备一定美感的产品。这种选择是我们内心中某些心理历程的结果。这种存在我们平时很难觉察,然而所谓的潜意识能够将我们与某种产品联系在一起,从某种程度上"强迫"我们进行选择和购买。

产品的新颖设计能够有效提升其市场份额,从而提高销量,最终实现获利的目的。在设计过程开始之初,对产品的购买目标群体进行深入的研究和了解非常重要,设计需要根据这一调查和研究进行调整。平面设计在市场营销中扮演了非常关键的角色。大多数消费者通常能够接触到具有宣传目的的平面设计,包括海报、公告栏、宣传册、目录、宣传单等。除了这些平面设计形态,人们还会接触到其他一些设计形式,如杂志、包装、公司的视觉识别、图书封面……

设计的第一步是一个版式中元素的布局设计。版式是照片、插画和文字以一种独特的方式进行放置,从而使信息的接收者能够很容易地获取信息。运用在一个特别版式下的信息量没有固定的限制,可以很简单也可以很丰富。平面设计师的任务多种多样,可以是异常艰巨、复杂,亦可以非常简单,前者要求一个专业设计师或整个创意团队具有良好的工艺水准和机灵、敏锐的特质以具备妥善处理问题的能力。

版式的设计优势和信息量一样复杂,一个良好的版式设计能够帮助信息接受者顺利地获取想要了解的信息。通常,合理的版式可以同时应用到印刷品和电子媒介之上,并且需要图案的辅助。

除了一个功能性解决方案,设计师还需要在应对挑战的同时对美学程度进行关注。在元素的配置过程中,一个相当关键的要素是先确定设计材料的独有视觉特性,然后确认版面中的特定图案,并使图案在产品的配色方案中实现统一。通过一个清晰的线条网络,促进版式中元素的精确和快速定位。一方面,线条网络可以使元素间的定位更加自由;而另一方面,它可能会约束或稍微对外观进行协调,从而改变呆板的布局模式。设计中运用的微妙笔触,是设计团队经验与智慧的写照,也正是它们决定了平庸与非凡之间的距离。

合理的版式设计方案可以通过直觉来设计,然而,在平面设计和综合性设计领域工作多年之后,即会建立起轻松获得信息的基础以及高度审美感。答案是黄金分割和斐波纳契数列。毫无疑问,拥有最佳设计感的产品,具有高度的美学理念,在斐波纳契数列中占有一定比例。黄金分割象征着两个物体间的比例、文本或者某种实体的尺寸。这个术语可以用一条线的分割来进行描述。举例说明,一条线被分成不等的两个部分,然而较短的部分和较长的部分之间的比例与较长的部分与整条线之间的比例相同。拥有这种流畅比例的元素布局能够令人赏心悦目。

斐波纳契数列因数学家斐波纳契提出而得名,该数学家来自于意大利比萨城,他在12世纪末提出了这一理论,而这一发现至今在科学和艺术领域仍是一个亮点。

以使信息展示更加简化为出发点,版式中元素定位的一个重要事实是人们首先获取照片和插画,然后再将注意力转移到文本之上。

# PREFACE

and design in general, after many years, the basis of easier receiving of information and highly aesthetic quality was established. The answer was the golden section and Fibonacci numbers. There is no doubt that best designed products, with high level of aesthetics have the proportions that are to be found in Fibonacci numbers. The golden section represents the ratio between two objects, sizes of the texts or certain entities. This term is best to be described through the division of a line. Namely, a line is divided into two unequal parts, but the ratio between the shorter part and the longer one is the same when compared with the ratio between the longer part
of the line and the line as a whole. Elements that are characterised by such a smooth proportion are pleasing to the human eyes.

Fibonacci numbers were named after the mathematician Fibonacci, respectively Leonardo of Pisa – Italy, who created them at the end of the $12^{th}$ century, and whose discoveries are still in the spotlight in the realm of science and art.

Rather important fact for positioning of the elements on the formats, with the goal of presenting the pieces of information more easily, is that people first spot photographs and illustrations, and only later do they pay attention to the text.

Text is an element of high importance on a graphically designed format, and it is to be considered with a special care. The choice of typography is momentous part of designing, because the way of transmitting the written idea depends on the appearance of the font and formatting of the text.

Each font has its own features that help us evoke in and convey the desired emotion to the reader. Therefore we can for example use neutral, elegant or hand-written typography and despite the text being the same, diverse effects can be accomplished when using different letter types. There are two basic types of typography - serif typeface and sans-serif typeface. The difference between them is that serif typeface has an ornament (serif) at the ends, whereas sans-serif has none and it is suitable for those texts whose size ought to be quite small due to the volume of the content. When choosing the right type of the typography, it is crucial to adjust the design of it, as well as the message we desire to convey to the aimed group. This can help us when selecting the letters and later on make the communication easier. The very composition of the words into blocks is a peculiar case, where it is necessary to see to it that the block is readable. The established rules that make this job easier refer to the length of the lines and the space between them. It is said that the optimal length of the line text should be between 50 and 70 characters, and the ratio between the very lines should dwell in the realms of the golden spiral or Fibonacci number.

On one hand, the basis for a solid commercial usage of design is the sound knowledge of the necessities of those people whom the design is intended for, and on the other hand, we need to have a clear image of what we want the design to state. By having a clear insight into both of these sides, and using the craft of graphic design we are able to provide illustrious solutions.

## Igor Milanovic & Ijija Dragisic (Serbia)

Translation for preface from Serbian to English: Jelena Grubetić and Stefan Vučićević.

# 前言

　　文本元素对平面设计模式来说非常重要，因此，它常常需要进行特殊的"对待"。字体的选择是设计的重要组成部分，这是因为书面意念的传递方式取决于字体的外观和文本的格式。

　　每个字体都有其自身的特点，有助于使设计师的设计情感与读者产生共鸣。因此我们可以利用中性、优雅或手写字体，即使文本相同，不同的文字类型也能够产生丰富的视觉效果。有两种基本的字体，即衬线字体和无衬线字体。二者之间的差别是，衬线字体在末端有点缀（衬线），而无衬线字体则没有，它适用于受内容限制的小尺寸文本。在选择正确的字体类型的同时，我们更需注意对设计和要传达给目标群体的信息进行调整。这可以帮助我们进行文字的选择，从而使传达变得更加简便、易懂。

　　单词构成模块是一个独特的情况，有必要注意的是，模块是值得一读的。使这项工作更加容易的既定规则涉及线条的长度以及它们之间的距离。据说，最优化的文本线条长度应当介于50到70个字符之间，而线条之间的比例也应处于黄金螺旋或斐波纳契数列范围内。

　　一方面，一个坚实的商业用途设计的基础是对设计的目标群体具有深入的了解，洞悉他们内心所需；而另一方面，我们需要拥有一个清晰的理念，也就是说需要明白设计想要表达的实质。在这两个方面的基础上，运用清晰的思维和洞察力，借助于独特的平面设计技巧，我们就自然能够提出最佳的解决方案。

伊戈尔·米拉诺维奇、伊利亚·德拉吉斯克（塞尔维亚）

前言翻译（塞尔维亚语译成英语）：伊莲娜·格鲁贝迪克，斯特凡·乌希赛维克

# Cascades Annual Report on Sustainable Development 喀斯喀特公司年度可持续发展报告

Design Agency:
**Paprika**
Creative Director:
**Louis Gagnon**
Designer:
**Daniel Robitaille**
Client:
**Cascades**
Nationality:
**Canada**

设计机构:
红辣椒设计工作室
创意总监:
路易斯·盖格农
设计师:
丹尼尔·罗比泰勒
客户:
喀斯喀特公司
国家:
加拿大

The designers had to create a document announcing Cascades environmental results in each of the three spheres of sustainable development: environment, social and economy.

They conceived a document entirely made of recyclable materials; the cover of the report is made with a SanizorbMC pig mat, manufactured 100% from recycled materials, and the inside of the report is printed on Rolland ST100 paper, composed of 100% recycled fibres, certified FSC recycled, EcoLogo and Process Chlorine-free and manufactured in Quebec by Cascades using biogas. The document is not only recycled, but may be reused as a notebook. Cascades is thus driving back the recycling stage by giving the report an additional function.

设计师设计的这一年度报告旨在公布喀斯喀特公司在环境、社会和经济三个可持续发展领域的生态情况。

他们精心构思了一个全部由可循环再用材料构成的文本；报告的封面运用了SanizorbMC吸收材料，100%由可回收材料制成；报告的内文完全以罗兰ST100纸质为原料，该纸质材料由获得美国联邦科学委员会认证的100%可回收再利用纤维构成，而获得北美最具影响力的环境标准和认证标志－EcoLogo标签，则由喀斯喀特公司运用沼气，经无氯气处理，在加拿大魁北克制造。

该文献不仅可以回收再利用，还可以作为笔记本重复使用。这一额外功能的添加成功地帮助喀斯喀特公司重新占领了循环利用的舞台。

# Mendes Júnior Annual Report 2011

2011曼德斯朱尼奥尔公司年度报告

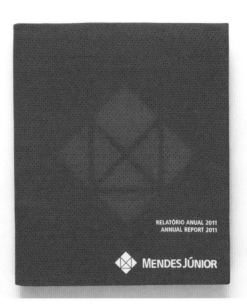

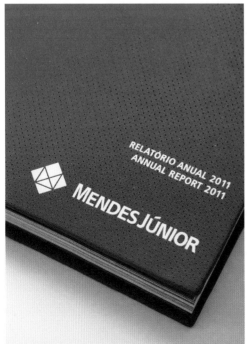

Design Agency: **Greco Design**
Designer: **Tidé, Diego Belo, Victor Fernandes, Alexandre Fonseca**
Client: **Mendes Júnior**
Nationality: **Brasil**

设计机构：Greco设计公司
设计师：蒂德、迪亚戈·贝罗、维克多·费尔南德斯、亚历山大·丰塞卡
客户：曼德斯朱尼奥尔公司
国家：巴西

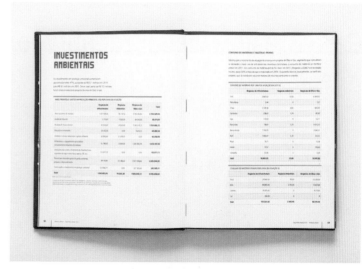

The report summarizes Mendes Júnior's path taken during 2011, while also presenting the institution's position for the future. The concept of "expanding limits" spells out, as a metaphor, Mendes Júnior's activities. The malleable cover was made of EVA and lined with the same fabric that is used in the employees' uniforms. The graphics project is based on a flexible and dynamic grid and the titles are made of a font type with an exclusive design.

这份年度报告总结了曼德斯朱尼奥尔公司在2011年的工作，同时也展望了公司未来的发展方向。德斯朱尼奥尔公司将"超越极限"作为发展的口号。报告的可塑性封面由EVA设计，与公司员工制服采用了相同的纺织材料。报告内的图表以动态网格为基础，所有的标题都采用了独家设计的字体。

# Care Brasil – Visual Identity for Annual Reports

"关注巴西" – 年度报告视觉识别设计

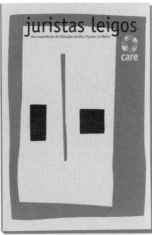

Design Agency:
**CASA REX**
Creative Director:
**Gustavo Piqueira**
Designer:
**Gustavo Piqueira / Samia Jacintho**
Client:
**Care Brasil**
Nationality:
**Brazil**

设计机构：
CASA REX设计工作室
创意总监：
古斯塔沃·皮奎拉
设计师：
古斯塔沃·皮奎拉，萨米亚·杰西罗
客户：
关注巴西机构
国家：
巴西

For many years, NGO promotional brochures closely followed the very corporate style of the business world in order to attract the right kind of serious investor. For some of them, this initial affirmation process has now been concluded, allowing for increased creative scope and flexibility. This was the starting point of the visual concepts for the publications that portray the projects developed by Care Brasil. Working within the geometry of graphic structures, the designers created diverse and irregular shapes that appear in different ways throughout the pages, reminding us that the end result of an NGO isn't about numbers. It is about humanity.

多年来,非政府组织宣传手册为了吸引更多真正投资者的目光,遵循了商业界的企业化设计风格。在他们中的一些人看来,这种最初的加工方式并不一定要一成不变,允许创意氛围的延伸,从而使其具有更强的灵活性。这是以描绘"关注巴西"开发方案为主题的系列出版物视觉理念的出发点。设计师运用几何形态平面结构,精心打造了多样化的不规则图形,以各种方式出现在页面之中,使读者自然领会到非政府组织的最终目标并不是抽象的数字,而是人文主义精神。

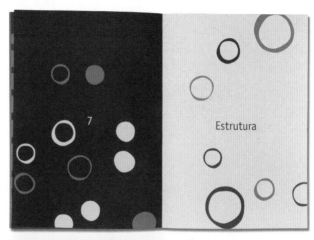

Meus colegas
prestem atenção
no que aqui vou falar
Dos executivos
em geral
E com eles venho
executar
É bom que fiquemos
atentos para não mais
nos enganar
Depois de conhecer
o dever do executivo,
queremos lhes dizer
Que há tantas
coisas erradas
Que venha acontecer
Todos sendo
enganados
Com os olhos vedados
sem ver...

# Alliance Resource Partners & Holdings Annual Report
联盟能源合作伙伴控股公司年度报告

It is Annual Reports for Alliance Resource Partners and Alliance Resource Holdings. Client wanted both Annual Reports to showcase the shift in corporate philosophy to a cleaner, safer and more environmentally safer coal plant. From the uncoated, typographic covers to the oversaturated, outdoor photography - both pieces were a pleasant and unexpected departure from the industry norm.

该项目是专为联盟能源合作伙伴及控股公司而设计的年度报告。客户希望年度报告能够完美展现出企业向更洁净、更安全和更环保的燃煤火电站转变的哲学理念。从无涂层印刷封面到过饱和户外摄影，处处流露出愉悦、独特的气息，打破了原有工业设计的模式。

| | |
|---|---|
| Design Agency: | 设计机构： |
| Walsh | 沃尔什设计工作室 |
| Creative Director: | 创意总监： |
| Kerry Walsh | 克里·沃尔什 |
| Designer: | 设计师： |
| Chad Mjos | 扎德·莫杰斯 |
| Client: | 客户： |
| Alliance Resource Partners | 联盟能源合作伙伴控股公司 |
| Nationality: | 国家： |
| USA | 美国 |

# Soul Hair Annual Report 发艺空间年度报告

Designer:
**Freddy Thomas**
Client:
**Soul Hair**
Nationality:
**UK**

设计师:
**福瑞迪·托马斯**
客户:
**发艺空间**
国家:
**英国**

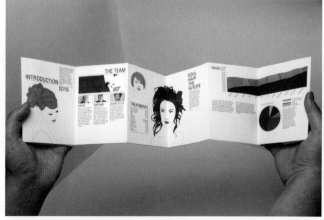

It is the annual report design for the Soul Hair Company in Leicester. The design focuses on the hair of the illustrations to highlight the nature of the business. It also is a pocket - size concertina fold, which fits the hand bag size style of many hair fashion magazines. The back of the report also features a promotional poster with the companies slogan "Make them all envy you", which features typography inspired by hair.

该项目是专为坐落在英国莱切斯特市的发艺空间公司而设计的年度报告。设计以发型插画为主题，旨在突出该公司的特色所在。口袋图书大小的风琴式折叠工艺能够与很多发型设计杂志的手提袋尺寸相匹配。该报告的后身还附设了一个带有该公司宣传语"让你成为被羡慕的焦点"的宣传海报。字体的设计受到了头发的启发。

# American Foundation for the Blind Annual Report 美国盲人基金会年度报告

AFB's annual report needed to address the long-standing work that this organisation has accomplished and continues to implement while also asking people to look forward to the future, together. The report is printed as two halves, "The Present" and "The Future", which come together under the support and dedication of AFB's network.

美国盲人基金会年度报告旨在彰显该组织长期以来所获得的工作成果以及未来不断完善的规划蓝图，同时号召人们一同憧憬美好的未来。该报告分两部分印刷，它们分别是"今天"和"明天"，二者通过美国盲人基金会的网络支持和共享而结合在一起。

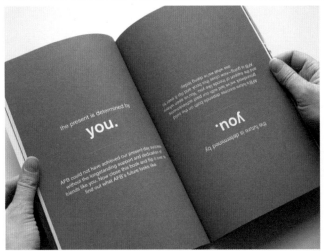

Design Agency:
**TODA**
Creative Director:
**Marcos Chavez**
Designer:
**V. Ann Suvarnapunya**
Client:
**American Foundation for the Blind (AFB)**
Nationality:
**USA**

设计机构：
TODA设计工作室
创意总监：
马科斯·查韦斯
设计师：
V.安·苏瓦纳普亚
客户：
美国盲人基金会(AFB)
国家：
美国

# Endeavor Impact Report

Endeavor 非营利组织年影响报告

Design Agency:
**TODA**
Creative Director:
**Marcos Chavez**
Designer:
**V. Ann Suvarnapunya**
Client:
**Endeavor Global**
Nationality:
**USA**

设计机构：
TODA设计工作室
创意总监：
马科斯·查韦斯
设计师：
V.安·苏瓦纳普亚
客户：
Endeavor 非营利国际组织
国家：
美国

Endeavor is a global not-for-profit organisation working to promote entrepreneurs in emerging markets. In 2007, the organisation celebrated a decade of operation and wanted a bright, bold annual report in commemoration of this milestone.

Endeavor是一家非营利组织，该组织致力于与新兴市场的高影响力创业者进行合作，帮助其改善公司并实现公司增长。2007年，该组织为庆祝其创办十周年，特别委托TODA设计工作室为其打造一个醒目、大胆的年度报告，以纪念这一里程碑时刻。

# Endeavor Impact Report

Endeavor 非营利组织年影响报告

Design Agency:
**TODA**
Creative Director:
**Marcos Chavez**
Designer:
**V. Ann Suvarnapunya**
Client:
**Endeavor Global**
Nationality:
**USA**

设计机构:
TODA设计工作室
创意总监:
马科斯·查韦斯
设计师:
V.安·苏瓦纳普亚
客户:
Endeavor 非营利国际组织
国家:
美国

A single sheet of white paper is a visual metaphor representing the journey of an entrepreneur. It is the transformation of an idea that grows and takes flight. A series of photographs brings this concept to life, showing growth, networking and community.

这个单页白皮书是一个视觉象征，寓意一个企业家的旅程。正是因为思想的不断转变促进了更大的发展和进步。一系列的照片将这一理念渲染得分外生动，完美展现了发展、协同合作以及团体的互助精神。

# AUSL Valle d'Aosta Annual Report
意大利莱·达奥斯塔大区年度报告

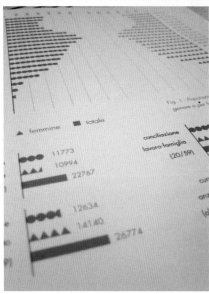

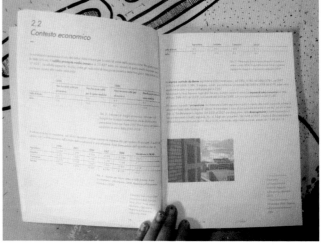

| Design Agency: | 设计机构: |
| --- | --- |
| **LLdesign** | LLdesign设计工作室 |
| Creative Director: | 创意总监: |
| **Lorella Pierdicca** | 洛莱拉·佩蒂卡 |
| Designer: | 设计师: |
| **Lorella Pierdicca** | 洛莱拉·佩蒂卡 |
| Client: | 客户: |
| **AUSL Valle d'Aosta** | 意大利莱·达奥斯塔大区 |
| Nationality: | 国家: |
| **Italy** | 意大利 |

The cover image represents a growth path, a landscape around which they settle people. The citizen is at the centre, and on the basis of their needs are being built up actions of the Government. The purpose of the annual report is to assess the results of actions made through the feedback of citizenship.

封面图案代表着发展的曲线图以及环绕在人们周围的景观。公民被设置在核心地带,政府的措施建立在公民的需求基础之上。该年度报告的目的是通过公民的反馈对措施的结果进行评估。

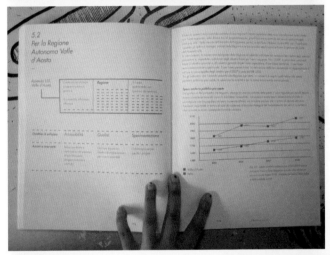

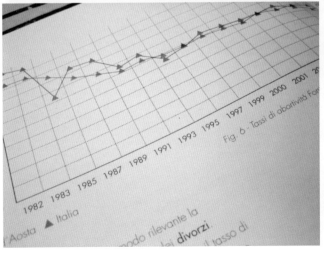

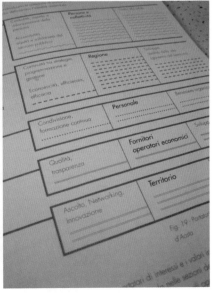

# Meteoric Resources Annual Report
大气资源公司年度报告

Design Agency:
**key2design**
Designer:
**Gemma Wilson**
Client:
**Meteoric Resources**
Nationality:
**Australia**

设计机构:
key2design设计工作室
设计师:
杰玛·威尔逊
客户:
大气资源公司
国家:
澳大利亚

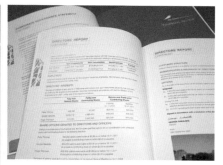

Meteoric Resources are an Australian mining company based in Western Australia, the design for the 2007 Annual Report conveys an industrial modern feel with the use of simplified graphics and modern typefaces, this allows for eligibly, yet creates reader stimulation throughout the document.

　　大气资源公司是澳大利亚一家矿业公司,坐落于澳大利亚的西部,这个2007年度报告设计巧妙运用简约的图案和现代字体,完美传达了一个工业化的现代美感,恰如其分的设计令读者在阅读文本的同时深受感染和鼓舞。

# Strategic Vision & Business Plan

战略眼光与商业规划

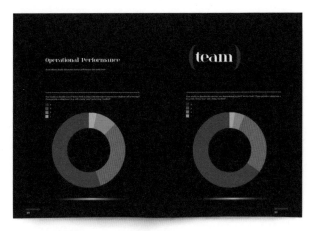

Design Agency:
**www.ewencom.com**
Creative Director:
**Martin Ewen**
Designer:
**Paul Sewell**
Client:
**North Mersey Health Inormatics Service**
Photography:
**Paul Sewell**
Nationality:
**UK**

设计机构:
www.ewencom.com设计工作室
创意总监:
马丁·伊文
设计师:
保罗·史威尔
客户:
北默西河健康信息服务中心
摄影:
保罗·史威尔
国家:
英国

Developed for North Mersey Health Informatics, this Strategic Business Plan was designed using only two colours and features photography of the region along with specially shot images of key personal and clients of the organisation.

该项目是专为北默西河健康信息服务中心开发的战略商业规划,仅仅运用两种色调和相关地点照片以及主要人物和客户的特写照片作为设计的主要元素。

# ASI Annual Report
加州理工州立大学学生协会年度报告

Design Agency:
Gas Creative Group
Creative Director:
Andrea Tinchinda / Kenny Flores
Designer:
Martin Lou
Client:
ASI (Associated Students Incorporated, Cal Poly Pomona)
Photography:
Eric Catig
Nationality:
China / USA

设计机构：
气体创意团队
创意总监：
安德里亚·缇辛达，肯尼·弗洛雷斯
设计师：
马丁·卢
客户：
加州理工州立大学学生协会
摄影：
埃里克·凯蒂格
国家：
中国，美国

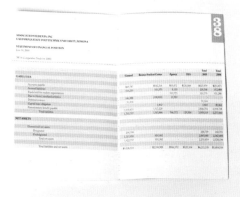

It is ASI's annual review that celebrates the achievements during the previous full academic year. The Annual Report highlights new and on-going strategic plans, audit reports, partnerships, sponsorships, and features faculty and department achievements.

该项目是加州理工州立大学学生协会的年度报告,旨在对过去的整个学术年所获得的收获进行总结和表彰。该报告突出了最新以及正在开展的战略规划、审计报告、伙伴关系、赞助商,并重点体现了成员和部门所取得的成就。

## NSF Annual Report 北极星基金会年度报告

Design Agency: **Hyperakt**
Creative Director: **Deroy Peraza / Julia Vakser**
Designer: **Julia Vakser / Jason Lynch**
Client: **North Star Fund**
Nationality: UK

设计机构: Hyperakt设计工作室
创意总监: 德罗伊·佩拉扎, 朱莉娅·瓦可塞
设计师: 朱莉娅·瓦可塞, 杰森·林奇
客户: 北极星基金会
国家: 英国

The concept for North Star Fund's annual report was to illustrate the building blocks of a just and equitable NYC. The report features individuals and organisations who contribute a valuable service to the city and surrounding areas. Their contributions are represented by icons that combine with others to form a greater image. The images are illustrations of iconic NYC landmarks.

北极星基金会年度报告的设计理念是阐释这个坐落于美国纽约的组织的公平与公正。该报告以对城市和周边地区做出杰出贡献的个人和组织为主题。设计师巧妙地运用图标将这些人所做出的杰出成就一一展示出来，并使图标之间建立起微妙的联系，从而形成了一个大幅的图像。图片中所突出的部分是纽约城市地标。

# Mersey Care Annual Report

默西河保健中心年度报告

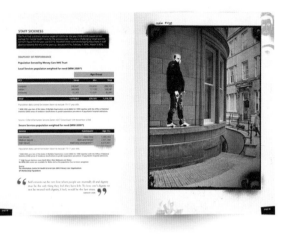

| Design Agency: | 设计机构: |
|---|---|
| www.ewencom.com | www.ewencom.com设计工作室 |
| Creative Director: | 创意总监: |
| Martin Ewen | 马丁·伊文 |
| Designer: | 设计师: |
| Paul Sewell | 保罗·史威尔 |
| Client: | 客户: |
| Mersey Care NHS Trust | 默西河保健中心 |
| Photography: | 摄影: |
| Paul Sewell / Martin ewen | 保罗·史威尔,马丁·伊文 |
| Nationality: | 国家: |
| UK | 英国 |

This report that titled "Right on" was themed on human rights values of fairness, respect, equality, dignity and autonomy. The design uses street art spray as the theme and features a range of case studies on service users and carers all of whom were created as spray paint stencils depicted on landmarks and suitable backdrops throughout the Liverpool city centre. It is the winner of 2010 North West CIPR Pride Gold Award for Best Publication.

该报告就人权问题上的公平价值、尊重、平等、尊严和自治提出"完全正确"的理念。该设计以街头艺术喷绘为主题,并以一系列针对服务用户和护工所展开的案例分析为特色。这些人物作为喷绘对象,与地标一同为利物浦市中心打造了一道别具一格的城市景观。该项目在2010年获得了英国西北部知识产权委员会颁发的最佳出版奖。

# Oxford Health NHS Annual Report
牛津国民健康保险信托基金会年度报告

Design Agency:
www.ewencom.com
Creative Director:
Martin Ewen
Designer:
Paul Sewell
Client:
Oxford Health NHS Foundation Trust
Photography:
Paul Sewell / Martin ewen
Nationality:
UK

设计机构：
www.ewencom.com设计工作室
创意总监：
马丁·伊文
设计师：
保罗·史威尔
客户：
牛津国民健康保险信托基金会
摄影：
保罗·史威尔，马丁·伊文
国家：
英国

This report was designed to look like an artists note book with a series of faces from the main ID reproduced as illustrations. The art style was extended further and used to create the theme for all the graphs and tables and the report was produced on a matt recycled paper to add to the authenticity of the design concept.

这份报告的设计乍看起来犹如艺术家的一个笔记本，即将一系列来自主体身份的头像进行复制，最后形成插画。这种艺术风格在这里得到了进一步延伸，旨在为所有图形和表格创造出一个主题，而该报告以一个亚光再利用纸张为原料，巧妙地为这一设计理念增添了纯粹之感。

# Robert Walters Annual Report 罗伯特·沃尔特有限公司年度报告

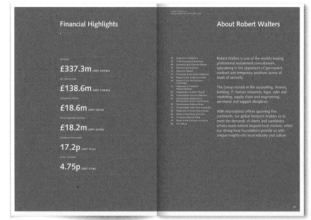

Design Agency:
**35 Communications (now Conran Design Group)**
Creative Director:
**Lee Hoddy**
Designer:
**Ryan Miglinczy / Chris Oates**
Client:
**Robert Walters**
Nationality:
**UK**

设计机构:
35 传达设计工作室(现更名为康兰设计集团)
创意总监:
李·霍迪
设计师:
瑞恩·米格林，克里斯·奥特斯
客户:
罗伯特·沃尔特有限公司
国家:
英国

It is the annual report that created for financial services recruitment agency, Robert Walters. In a tough financial climate Robert Walters' performance had been better than expected, yet they wanted to show the progress made in a respectful manner, using a limited colour pallete and reserved layout.

这一年终报告专为金融服务中介机构罗伯特·沃尔特有限公司而设计。在艰难的经济气候下，罗伯特·沃尔特有限公司优秀的运营成果着实令人惊讶，然而，他们依然希望该报告能够以低调的方式展现出他们所获得的进步，强调运用简单的配色方案以及保守的版式格局。

# RT Health Fund Annual Report

RT健康基金会年度报告

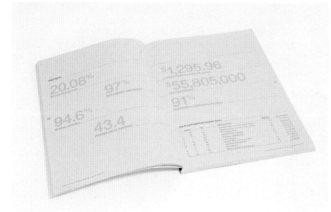

Design Agency:
**Team Scope**
Creative Director:
**Mark Burrough**
Designer:
**Fiona Boyan**
Client:
**RT Health Fund**
Nationality:
**Australia**

设计机构：
团队视野设计工作室
创意总监：
马克·博拉夫
设计师：
菲奥纳·博伊安
客户：
RT健康基金会
国家：
澳大利亚

Team Scope designed an attention-grabbing Annual Report with a punchy tagline that emphasised the report's theme and then backed it up with a clean easy-to-read interior with fun feature pages focused on the things that RT don't do.

团队视野设计工作室为RT健康基金会设计的年度报告醒目而独特，运用一个有力的标志性语言有效地突出了该报告的主题，并采用一个简洁、清晰、有趣的内页设计风格，突出了RT健康基金会尚未涉及的业务。

# John Holland Annual Review
约翰·霍兰德公司年度综论

Design Agency:
**Team Scope**
Creative Director:
**Mark Burrough**
Designer:
**Mark Burrough**
Client:
**John Holland**
Photography:
**Karl Schwerdtfeger**
Nationality:
**Australia**

设计机构:
团队视野设计工作室
创意总监:
马克·博拉夫
设计师:
马克·博拉夫
客户:
约翰·霍兰德公司
摄影:
卡尔·斯奇沃尔特菲尔德
国家:
澳大利亚

When John Holland asked Team Scope to design their annual review, the designers wanted to create a bold and dynamic document that would really celebrate the business's successes and achievements for the year. The designers created a magazine-style layout that gives the review a fresh, contemporary feel.

受约翰·霍兰德公司的委托，团队视野设计工作室为其提供年度回顾报告的设计方案。设计师意在创建一个大胆而充满活力的文本，使之真正体现出该公司在过去一年中所获得的成就和业绩。设计师巧妙地构建了一个杂志风格版式，为该报告营造出清新、时尚之感。

# RT Health Fund Annual Report
RT健康基金会年度报告

Design Agency:
**Team Scope**
Creative Director:
**Mark Burrough**
Designer:
**Mark Burrough**
Client:
**RT Health Fund**
Photography:
**Karl Schwerdtfeger**
Nationality:
**Australia**

设计机构:
团队视野设计工作室
创意总监:
马克・博拉夫
设计师:
马克・博拉夫
客户:
RT健康基金会
摄影:
卡尔・斯奇沃尔特菲尔德
国家:
澳大利亚

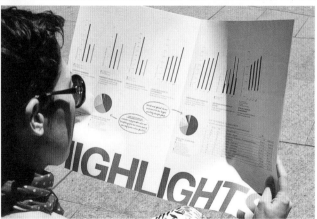

Team Scope was proud to play a part in history with the production of RT Health Fund's 119th annual report and this year the project was bigger than ever! The A3 sized report was designed to tell a story about the fund's big year and industry-leading levels of growth.

团队视野设计工作室很高兴能够在RT健康基金会第119届年度报告设计中扮演重要的角色,并且这一年的项目规模之大要超过以往任何一年。这个A3格式的报告讲述了该基金会在过去一年中所发生的大事记以及业界的领先发展水平。

# Croatian Design Annual 克罗地亚设计师协会年度报告

Design Agency:
**Sensus Design Factory Zagreb**
Creative Director:
**Nedjeljko Spoljar**
Designer:
**Nedjeljko Spoljar / Kristina Spoljar**
Client:
**HDD - Croatian Designers Society**
Photography:
**Nedjeljko Spoljar / Dejan Dragosavac Ruta**
Nationality:
**Croatia**

设计机构:
萨格勒布Sensus设计工厂
创意总监:
耐蒂艾尔克·斯颇里扎
设计师:
耐蒂艾尔克·斯颇里扎,克莉斯汀娜·斯颇里扎
客户:
HDD –克罗地亚设计师协会
摄影:
耐蒂艾尔克·斯颇里扎,德安·德拉格萨瓦克·卢塔
国家:
克罗地亚

Croatian Design Annual Book is accompanying with the annual exhibition of best Croatian graphic, product and interactive design. Using of different printing techniques, optical illusions and stickers makes it quite fun and functional to read and use. UV-varnished cover has been constructed to serve also as the bookmark, the content page and the book jacket.

该项目是专为克罗地亚设计师协会而设计的年度书刊、最佳克罗地亚平面作品、成果展以及互动设计。设计师运用多种印刷技巧、光幻觉以及贴纸令阅读和使用过程更加有趣和实用。紫外线亮漆封面同时与书签、图书内页以及护封完美结合在一起。

# Annual Report of "SIBUR" Company - Chemical Fertilisers

"SIBUR" 化学肥料公司年度报告

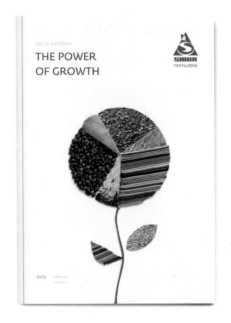

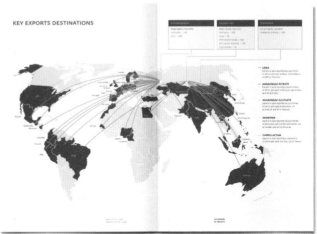

Design Agency:
**YellowDog**
Creative Director:
**Ilya Mitroshin**
Designer:
**Andrew Gorkovenko**
Client:
**"SIBUR" Company**
Nationality:
**Russia**

设计机构:
YellowDog设计工作室
创意总监:
伊利亚·米特罗辛
设计师:
安德鲁·戈科文克
客户:
"SIBUR" 化学肥料公司
国家:
俄罗斯

Annual report of "SIBUR" Company - chemical fertilisers Goods of "SIBUR" - is presented in different regions of the world. Prospects of the company's business are shown through traditional agricultural cultures because nitrogen fertilisers are one of the most important elements of agricultural production. Agricultural goods' images illustrate the database connected with countries where the company is presented. Images of grains associated with particular country and business direction are shown in the report.

该项目是专为"SIBUR"化学肥料公司而设计的年度报告方案,设计师巧妙地运用地图的形式将该公司生产的产品完美展现。由于氮肥是农业生产中一个非常重要的元素,因此,设计师运用传统的农业文化将公司的远景规划充分彰显出来。设计师运用农业商品图像举例阐释了该公司产品在各出口国家的销售情况。另外,报告中还大量运用了谷物图案,以与特定的国家和业务导向建立起微妙的联系。

# Annual Report of Novolipeck' Steel Plant

Novolipeck钢铁厂年度报告

Design Agency:
**YellowDog**
Creative Director:
**Ilya Mitroshin**
Designer:
**Andrew Gorkovenko**
Client:
**Novolipeck steel**
Nationality:
**Russia**

设计机构：
YellowDog设计工作室
创意总监：
伊利亚·米特罗辛
设计师：
安德鲁·戈科文克
客户：
Novolipeck钢铁厂
国家：
俄罗斯

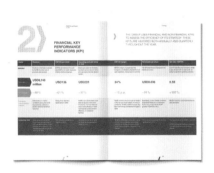

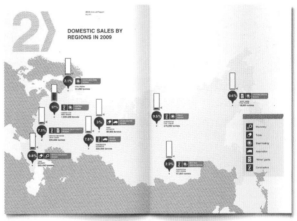
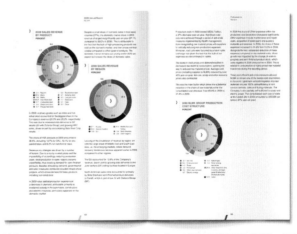

It is the annual report of Novolipeck' Steel Plant and the history of innovations (innovation in steel). The steel plant of Novolipeck is one of the world's leading steel producers. Historical documentation, archive photos underlining the role of Novolipeck' Steel Plant innovations in relative manufacturing, economics and techniques are presented in this report. The format of graphics and diagrams derives from the plant's production elements such as mill or finished goods. Colour spectrum simulates the spectrum of metal melting – from dark red to emerald.

该项目是专为Novolipeck钢铁厂而设计的年度报告，旨在将钢铁业的改革历史娓娓道来。Novolipeck钢铁厂是世界上领先的钢铁生产商之一。在该报告中，设计师巧妙运用历史文献、存档照片突出Novolipeck钢铁厂的革新在相关制造业、经济和技术领域的重要影响。图形和图表的格式取材自该工厂的产品元素，诸如搅拌机或成品等。色调以深红色向翡翠色过渡为主，与钢铁融化现象相得益彰。

# Air Canada Annual Report 加拿大航空公司年度报告

Design Agency:
**Jordan Puopolo Design**
Creative Director:
**Jordan Puopolo**
Designer:
**Jordan Puopolo**
Photography:
**Jordan Puopolo**
Nationality:
**Canada**

设计机构：
乔丹·普坡罗设计工作室
创意总监：
乔丹·普坡罗
设计师：
乔丹·普坡罗
摄影：
乔丹·普坡罗
国家：
加拿大

It is a self-inspired piece celebrating the 70th anniversary of Air Canada. The book celebrates the history of the airline through the use of vintage imagery and illustrations that were inspired by Air Canada ads from the past.

该项目是专为加拿大航空公司成立70周年而提供的个人设计方案。这本书通过运用大量的复古图像和插画阐释了该航空公司悠久的历史，而这些图像和插画的设计受到了加拿大航空公司广告演变的启发。

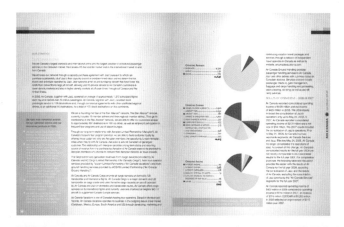

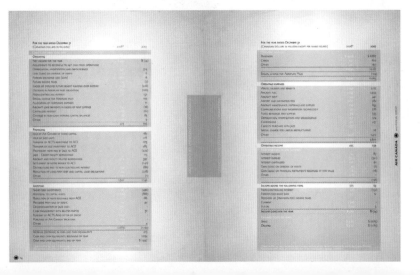

# Big Annual Review 纸箱公民剧团年度回顾

Design Agency:
**Interabang**
Creative Director:
**Adam Giles / Ian McLean**
Designer:
**Adam Giles / Ian McLean**
Client:
**Cardboard Citizens**
Photography:
**Various**
Nationality:
**UK**

设计机构：
Interabang设计工作室
创意总监：
亚当·贾尔斯，伊恩·麦克林恩
设计师：
亚当·贾尔斯，伊恩·麦克林恩
客户：
纸箱公民剧团
摄影：
来自各种摄影渠道
国家：
英国

The Review adopts a "big" theme to cover Cardboard Citizen's massive year and highlight their huge achievements. This theme is also realised in the format and the tightly cropped imagery, with the loose binding suggesting the language of newspapers and important issues. A small format, tipped-in donation form points towards a little help going a long way.

该报告以"宏大"为主题，旨在对纸箱公民剧团收获颇丰的一年进行完美回顾，同时突出该公司所获得的伟大成就。此外，这一主题也体现在报告的开本以及压缩图像之中，自由的装订方式暗示了报纸和重要发行物的设计语言。小开本的插页作为整个报告的亮点之一，向人们证明了个人的一点点爱心将会产生很深远的社会效益。

# SolarWorld Report 阳光世界股份公司年度报告

Design Agency: 设计机构:
**Strichpunkt Design** Strichpunkt设计工作室
Client: 客户:
**Solarworld AG** 阳光世界股份公司
Photography: 摄影:
**Sandra Schuck** 桑德拉·舒克
Nationality: 国家:
**Germany** 德国

The Group annual report shows how conviction and commitment to sustainable ecological energy supply result in long-term success and growth. Together, employees and customers work on the vision of a SolarWorld. The lavishly illustrated feature magazine portrays people who live out their commitment to a solar energy in different ways - for example a development projects in Malioder Haiti, environmentally conscious home owners in France or as employees of SolarWorld. The comprehensive reporting is enriched with compact fact sheets and information graphics on the energy market and the Company and by a flip book of the current TV commercial featuring Larry Hagman. The comprehensive sustainability reporting according to GRI was disincorporated into a separate print-on-demand report, which can be ordered per postcard.

这一集团年度报告完美展现了阳光世界股份公司长期致力于可持续性生态能量供给事业所获得的成功和发展,同时也展现了员工和客户共同努力下为该公司所打造的美好未来。丰富的插画专题杂志展示了以各种方式提倡有效利用太阳能的人物和活动,例如Malioder Haiti地区的开发项目以及法国拥有极强环保意识的房主或阳光世界股份公司中的员工等。紧凑的说明性内容、大量的有关能源市场和公司的信息图形以及以介绍赖瑞·海格曼为主题的当今影视广告分析图书使整个报告更加丰富、引人注目。这一系统化可持续性报告根据全球报告倡议组织的要求,被拆分成一个独立的限量版报告,可以利用明信片进行订购。

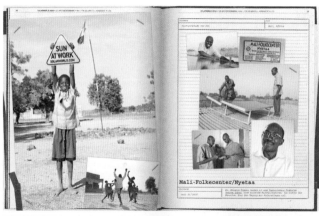

### Mali-Folkecenter/Nyetaa

**FRANKREICH:**
MIT SYSTEMKOMPETENZ NEUE DÄCHER UND MÄRKTE EROBERN

- Europäischer Solarmarkt mit Zukunft
- Beschleifiches Wachstum seit 2007 im klassischen "Ausstieg"
- Ästhetisch und technisch anspruchsvolle Kunden
- Besonders gute Förderung von selbstgenutzten Dachanlagen
- SolarWorld trifft auf seinem System Energieglück* die hohen Ansprüche von Markt und Kunden

„MIT EINER SOLARANLAGE AUF DEM DACH KÖNNEN AUCH DIE FRANZOSEN IHREN EIGENEN STROM ERZEUGEN UND EINE GUTE RENDITE ERZIELEN."

„SolarWorld: Der Name ist Programm. Die Idee ist genial einfach und funktioniert überall auf der Welt. Das trifft uns an."

„Wie schon ein römisches Sinnspruch sagt: Sol lucet omnibus – Die Sonne scheint für alle."

**A // Randall Richards**
Einst arbeitete er im Musikgeschäft in Hollywood, heute setzt er sich bei der SolarWorld für eine Zukunft ohne Öl ein. Dazwischen lagen berufliche Stationen als Pilotjunge und als Manager.

**B // Kevin Kilkelly**
Als President der US-Vertriebstochter ist er für den Markt von Kanada bis Chile verantwortlich. Anders als noch bei seinen Eltern und Großeltern ist unseren Kindern die Solarenergie schon ganz vertraut. Die Bedeutung für ihre Generation ist ihnen sehr bewusst.

**G // Angie Urrutia**
Eigentlich war die gebürtige El Salvadorianerin nur auf der Suche nach einem Job und fing eher zufällig in der Finanzabteilung in Camarillo an. Doch mit der Zeit wurde ihr bewusst: Dieser Job ist mehr wert als der Gehaltscheck.

**H // Melinda McCutchan-Davis**
Sie liebt alles, was sie in der Natur begeistert. Segeln, Goldschürfen oder Bergwandern, der Schutz des Planeten liegt ihr am Herzen. Als Technikerin für die Instandhaltung in Hillsboro setzt sie sich dafür ein.

**C // Jennifer Ryan**
Ein Kindheitserlebnis hat sie geprägt: Wie sie in der Oberstufe schaudernd auf dem Rücksitz von Mamas Toyota nach Benzin aussehen musste. Eine Motivation, heute für die Unabhängigkeit von Öl einzustehen.

**D // Brenda Betts**
Sie begann ihren Job als leitende Technikerin bei der SolarWorld 2008, als die neue Zellfabrik in Hillsboro fast noch leer stand. Dann kam der große Moment: Die erste Zelle lief vom Band. Heute sind es mehrere Millionen pro Monat.

**I // Jenny Yan**
Die Automationsingenieurin ist für eine reibungslose Modulfertigung in Camarillo verantwortlich. Qualität hat für sie Top-Priorität in der Produktion. Aus einem einfachen Grund: Ohne Qualität keine Kunden.

**J // Rusty Pittman**
Der US-Marketingchef kommt aus der Baubranche. Das weckt das Potenzial der Marke SolarWorld, und er sieht die große Chance, dass immer mehr Bauherren erkennen, dass eine Solaranlage ein kluges Investment ist.

**E // Max Tjipto**
Als junge Ichänner er den Traum vom Fliegen. Andres als gedacht, ist er nicht Flugzeugingenieur geworden, sondern sorgt heute als IT-Experte für ein effizientes Zusammenspiel der Produktion in Hillsboro und Camarillo.

**F // Ajay Garg**
Er wollte schon immer Ingenieur werden und hat das verwirklicht. In der Modulfertigung arbeitet er hart dafür, Kosten zu senken und die Qualität zu erhöhen. Ein ganz anderer Wunsch ging 2010 für ihn in Erfüllung: Sein Sohn kam zur Welt.

**K // Warlenia Bryant**
Die Ingenieurin ist eines der neuen Gesichter in der Modulfabrik in Hillsboro. Lange hat sie zuvor in der Automobilbranche gearbeitet. Für sie ist klar: Die Solartechnologie wird die nächste große Industrie.

**L // Michaela Ramsey**
Einst wollte sie Astronautin werden, heute konzentriert sie sich auf Umweltthemen. Die promovierte Ingenieurin verbessert den Kristallisationsverfahren. Sie will dazu beitragen, dass sich mehr Menschen für eine Solaranlage entscheiden.

# Gullkalven / The Golden Calf

Gullkalven 学生创意传达设计锦标赛年度报告 /金牛犊年度报告

Design Agency:
**Studio 3**
Creative Director:
**Paul van Brunschot**
Designer:
**Sindre Martin Dahl / Helene Hofstad Havro / Christian Horseng / Jon Koslung / Lene Sundsbø / Tore Teksle**
Photography:
**Sindre Martin Dahl**
Nationality:
**Norway**

设计机构:
Studio 3设计工作室
创意总监：
保罗·凡·布鲁恩斯科特
设计师：
辛德拉·马丁·达哈尔，海伦·霍夫斯塔德·哈维罗，克里斯汀·霍森，
琼恩·克斯朗，琳恩·萨德斯波，特勒·泰克斯勒
摄影：
辛德拉·马丁·达哈尔
国家：
挪威

It is the annual student championship in creative communication. Every year The Golden Calf changes its appearance. Traditionally it has always been an ox calf, but this time the designers went searching in the own neck of the woods for inspiration. The Golden Calf of 2010 was reincarnated as a deer calf and every student was invited to hunt for it and other trophies in the vast creative terrain.

该项目是专为挪威Gullkalven 学生创意传达设计锦标赛而设计的年度报告。每年，该锦标赛的形象更换一次。以往的形象一直以一个公牛为主题，然而，这一次，设计师自己深入到特定地点寻找设计灵感。该锦标赛的2010年度报告以幼鹿为新形象，每个学生都可以在广阔的创意地带发挥无限的创造力。

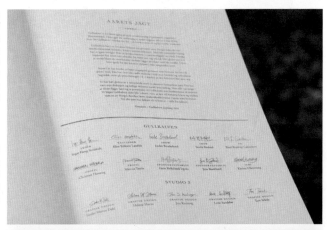

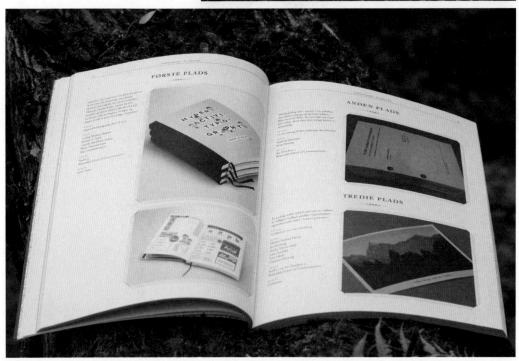

# Pliva Annual Report
普利瓦公司年度报告

Design Agency:
Sensus Design Factory Zagreb
Creative Director:
Neda Segovic / Nedjeljko Spoljar
Designer:
Nedjeljko Spoljar / Jadranko Marjanovic
Client:
Pliva d.d.
Photography:
Dag Sola Orsic / Damir Fabijanic / Getty Images
Nationality:
Croatia

设计机构:
萨格勒布胜塞斯设计工作室
创意总监:
奈达·赛格维克,耐德里克·斯普理查
设计师:
耐德里克·斯普理查,扎得兰科·玛伽诺威克
客户:
普利瓦公司
摄影:
戴格·索拉·奥赛克,戴梅尔·法比加尼克,盖蒂图片公司
国家:
克罗地亚

It is the annual report for Pliva, the biggest pharmaceutical company in Central and Eastern Europe. Cover contains a list of all PLIVA's branches worldwide. Perforations on the cover jacket reveal letters that make company's motto: "Dedicated to Health".

该项目是专为欧洲中部和东部最大的制药公司－普利瓦公司而设计的年度报告。该报告的封面上巧妙地设置了一张介绍普利瓦公司在全球的分支机构的列表，将该公司在全世界的影响力充分彰显。封套上镂空部分设置的字母，完美地彰显出该公司的经营理念，即"为健康事业而努力"。

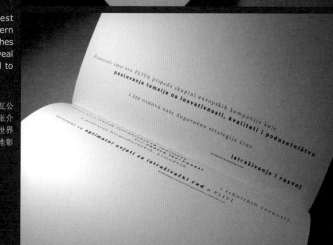

# Corporate Citizenship Report

"企业公民意识报告"

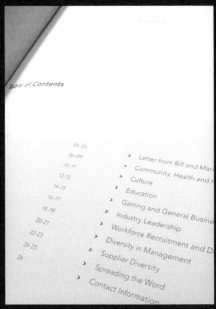

David Rengifo
Designer:
**Brad Towsley**
Client:
**Boyd Gaming**
Nationality:
**USA**

设计机构:
瑞恩视觉策略+设计工作室
创意总监:
大卫·莱吉弗
设计师:
布莱德·托斯里
客户:
波伊德博彩公司
国家:
美国

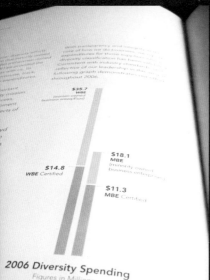

Boyd Gaming, a leader in the entertainment and gaming industry, is committed to making meaningful contributions to their community through four focused areas of charitable giving: Community Health and Human Services, Culture, Education, Gaming and general business associations. The overall design solution was to create a "hands on" approach demonstrating Boyd Gaming's charitable works.

波伊德博彩公司是娱乐和博彩产业的先驱,旨在通过四个慈善捐助的关键区为其团体提供有意义的捐助。这四个关键性慈善捐助区分别为社区卫生与公共事业、文化、教育、博彩以及一般商业协会。整个设计方案旨在创建一个"自己动手"的理念以展现波伊德博彩公司的慈善工作。

# SMAK1 Penabur / Yearbook SMAK1年鉴

Design Agency:
**83 Designhouse**
Designer:
**Almerio Adalberto Barros, Michella Virta**
Nationality:
**Indonesia**

设计机构:
83设计工作室
设计师:
阿尔莫里奥·阿达尔贝托、米凯拉·维尔塔
国家:
印度尼西亚

This is a project for one of the schools in Jakarta. This book contains so many memories they had during their time in school. It is usually known as Yearbook. The theme of the design is the colours of life. The designers summerise the students' life events in various coloured covers which are wrapped with a little touch of rolling or quilling papers. The word "Couleur" was adopted from French that means colour. Each and every tendril and line created are not without effort, they have to be measured, rolled and glued. So does our life. Each and everyone of us has our own colours in life. Every part of the memories are packed and taken a good care of, all in one elegant black board cover.

该项目是为雅加达一所学校所设计的年鉴，里面记录了许多学生的在校时光。设计的主题是生活的色彩。设计师在封面上利用卷纸和褶皱纸设计了多姿多彩的封面，并采用了法文"Couleur"（色彩）作为提名。每一条卷纸和线条的设计都经过精心的测量、卷制和黏贴。这一过程与我们的生活有相似之处。我们的生活都有不同的色彩，在年鉴中，每一份记忆都被妥善地保存好，外面套上了优雅的黑色硬板板外壳。

# Typokochbuch / book design and concept
诺维拉图书

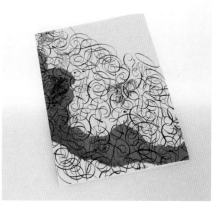

Design Agency:
sebastianhaeusler.design
Creative Director:
Sebastian Häusler
Designer:
sebastianhaeusler.design
Photography:
sebastianhaeusler.design
Nationality:
Germany

设计机构:
塞巴斯蒂安·豪斯勒设计工作室
创意总监:
塞巴斯蒂安·豪斯勒
设计师:
塞巴斯蒂安·豪斯勒设计工作室
摄影师:
塞巴斯蒂安·豪斯勒设计工作室
国家:
德国

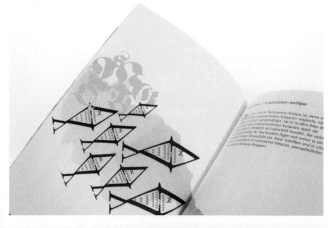

The typographic cook book is a small booklet that explains typographic principles in a fun way, using food and drink to aid the explanations, as a gray tone turns into lasagna and a font range transforms into fish. All illustrations are made entirely by means of typography and color.

　　这本印刷的食谱是一本通过一个有趣的方式来解释字体排版印刷学的小册子，通过使用食物和饮料作为辅助说明，将灰色调变成烤宽面条和将一种字体转化成鱼的形象。所有的插画都是通过使用排版印刷以及颜色完成的。

# Dwarse Vrouwen "女性面面观"印刷手册

Designer:
**Charlotte Aal / Karin ter Laak**
Client:
**Joke van der Zwaard**
Nationality:
**The Netherlands**

设计师:
夏洛特·奥尔,卡林特·拉克
客户:
约克·凡·德·扎瓦德
国家:
荷兰

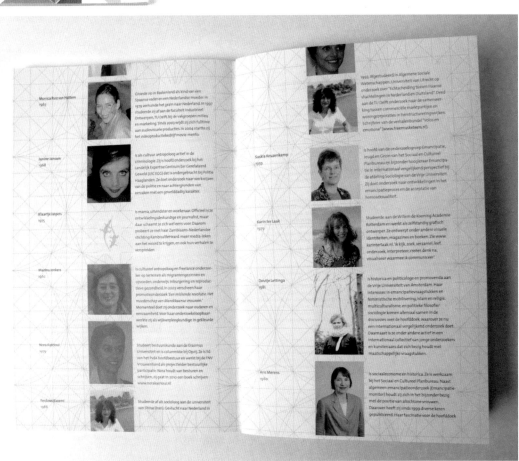

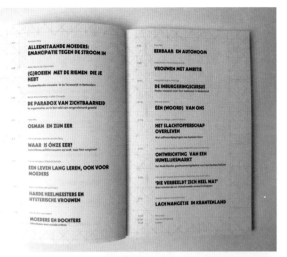

Tenacious but also connecting - the lines of thought sometimes diverge but on other levels they meet each other. The numerous lines stemming from patterns from the Orient, the base of many motives and colourful patterns, visualise the opinion of the author. Articles written by multiple authors are illustrated by a multitude of lines. The stories, columns, letters and the fairy tale form one extended continuous. They are interconnected and sometimes start "tenaciously" — perpendicular — on the page.

相对独立又确保相互衔接。有时思路会有偏差，但在某种层面上会出现交集。无数的线条从具有鲜明东方色彩的图案中衍生出来，许多线条和彩色图案形象地表达出作者的观点。众多作者编著的文章以大量的线条作为插画。故事、芬兰、字母以及神话故事形成了一个衔接自然的有机体。它们彼此之间相互关联，有时在页面上以垂直的形态呈现"执着"之态。

# Let's Be Active in Europe "让我们积极关注欧洲" 宣传册

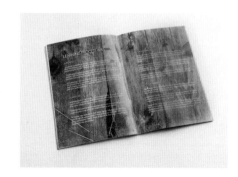

Design Agency:
Dimis design
Creative Director:
Igor Milanovic / Ijija Dragisic
Designer:
Igor Milanovic
Client:
NGO
Photography:
Nikola Andjelic
Nationality:
Serbia

设计机构：
Dimis设计工作室
创意总监：
伊戈尔·米拉诺维奇，
伊利亚·德拉吉斯克
设计师：
伊戈尔·米拉诺威克
客户：
非政府组织
摄影：
尼可拉·安迪里克
国家：
塞尔维亚

Let's be active in Europe is an NGO project of "The Centre for Non Violent Resistance". After finishing the project the designers had an additional task of collecting the thoughts and experiences of everybody involved. They designed the booklet and the web site in flash technology.

"让我们积极关注欧洲"是"非暴力抵抗中心"的一个非政府组织项目。在完成该项目的设计之后,设计师对该项目的相关人士的想法、经历进行征集。此外,设计师还运用动画编辑技术设计了广告宣传册以及网站。

# Enjoy the Countryside van het Verhaal

"欣赏乡间趣闻"印刷宣传册

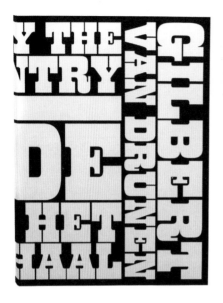

Designer:
**Roosje Klap**
Client:
**Gilbert van Drunen, Rotterdam**
Nationality:
**The Netherlands**

设计机构:
卢赛·克莱普设计工作室
客户:
鹿特丹吉尔伯特·凡·德伦内
国家:
荷兰

Gilbert van Drunen is an artist who works from a distinctive historical, almost political background and has created an oeuvre in installations, ceramics and drawings. He has collected his texts in this publication.

来自荷兰鹿特丹的艺术家吉尔伯特·凡·德伦内，拥有独特的历史和政治背景，在设备、制陶术和绘画领域均有杰出的创作。该出版物详细地阐释了与其相关的设计作品。

# Gde Etot Dom

"这样的房子在哪里？"宣传册

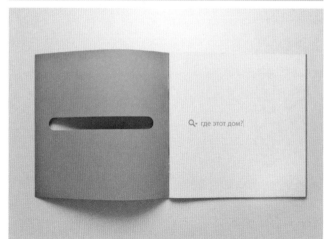

Design Agency:
Studio "ONY"
Creative Director:
Sergey Serezhin
Designer:
Russ Lobachev
Client:
Realty Information Portal "Gde Etot Dom"
Nationality:
Russia

设计机构：
"ONY"设计工作室
创意总监：
谢尔盖·塞拉辛
设计师：
拉斯·罗巴彻夫
客户：
"这样的房子在哪里？"房地产信息服务网站
国家：
俄罗斯

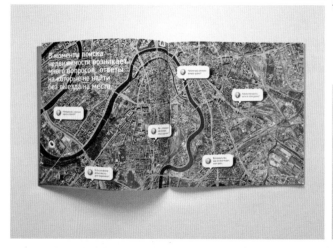

Name of project "Gde Etot Dom" can be translated to English as "Where is this house?" This web service searching realty allows to see photos of most houses and their environs in Moscow and other cities of Russia. This service allows seeing the house without needing to leave your chair. The booklet describes in detail the opportunity and benefits of this service.

名为"Gde Etot Dom"的项目在英语中可译成"这样的房子在哪里？"这一服务网站旨在为查找有关莫斯科和俄罗斯其他城市房产信息和相关照片的用户提供信息平台。用户通过该服务平台可以足不出户了解到房屋信息。该宣传手册详细地论述了该网站服务设计的用途与影响。

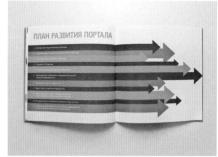

# Nike: Women's Training (Asia)

耐克: 女子训练系列运动鞋(亚洲)宣传册

Design Agency:
**PLAZM**
Art Director:
**Joshua Berger**
Designer:
**Thomas Bradley / Ian Lynam**
Client:
**Nike**
Photography:
**Allen Benedikt / Also known As / Anthony Georgis**
Nationality:
**USA**

设计机构:
PLAZM设计工作室
艺术总监:
约书亚·柏格
设计师:
托马斯·布拉德利,
伊恩·莱纳姆
客户:
耐克公司
摄影:
艾伦·贝内迪克特, Also known As公司, 安东尼·戈奥吉斯
国家:
美国

PLAZM created this work for the launch of the Nike's sisters training line. The designers were appealing to a younger, health-loving audience. The book also includes a CD with downloadable assets.

PLAZM设计工作室为耐克运动鞋女子训练系列的首发而提供宣传册设计。设计师以年轻、热爱运动的一族为设计目标。此外，该书还配有随书光盘下载。

# Who We Are, What We Do & How We Do It

"我们是谁，我们做什么，我们怎样做？" 宣传册

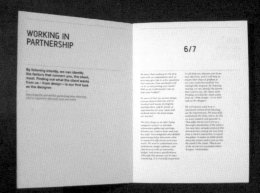

It is the booklet for promoting the skills and services of Bellamy Studio. Two-colour print on three colours of uncoated stock and stitched binding concealed by glued outer cover. Outer cover printed one colour on yellow card and die cut three positions. Cut on front continues the angle of the forward slash dividing the areas of expertise on the cover, and two cuts on the back for inclusion of two-colour foil block business card. It also includes fold out front cover revealing all areas of expertise and tipped in three-colour perforated and detachable postcard.

该手册的设计旨在充分展现贝拉米设计工作室的设计技巧和良好服务。三色无涂层原浆纸采用双色印刷，装订在一起后外覆以胶合封套。外层封套上的黄色卡片上方同时也印刷了与基底同样的色调，巧妙实现了三个层次的模切方式。最外层的模切结构表面运用斜杠将封面上的专业区域分隔开来；而其下方的第二层模切结构内囊括了双色商务名片模块。此外，折叠式封面充分展现了所有的专业领域，并粘附三色镂空可拆分明信片。

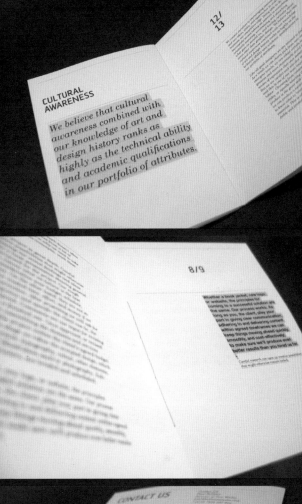

Design Agency:
**Bellamy Studio**
Creative Director:
**Andrew Bellamy**
Designer:
**Andrew Bellamy**
Photography:
**In house**
Nationality:
**USA**

设计机构:
贝拉米设计工作室
创意总监：
安德鲁·贝拉米
设计师：
安德鲁·贝拉米
摄影：
In house摄影工作室
国家：
美国

# Doku Manual 多库字体手册

Design Agency: **Tamer Köşeli**
Nationality: **Turkey**

设计机构: 泰默·科赛利设计工作室
国家: 土耳其

Doku is a monospaced, rounded, experimental display typeface, inspired from an organism. The typeface comes in 2 styles, each style contains 7 different weights. Doku is a suitable for posters, interfaces, logos and logotypes etc.
This manual designed for showing the capabilities of the Doku Typeface and direct the designer to use.

多库字体是一个等宽、圆润而丰满的试验性显示字体，该字体的设计灵感自一个有机体。多库字体拥有两种模式，每种模式都包含七个不同的重量。多库字体适用于海报、界面、商标以及标志等。
该字体手册的设计旨在展现多库字体的功能性并指导设计师如何使用。

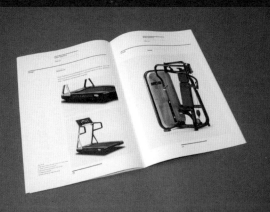
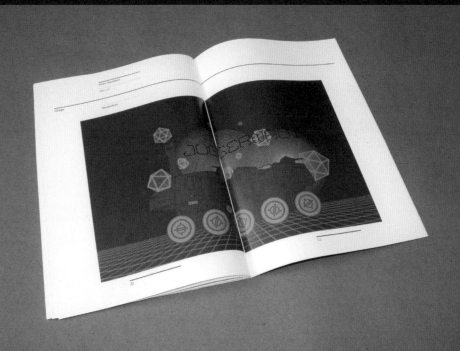

# Kopenhagen City Guide   哥本哈根城市向导手册

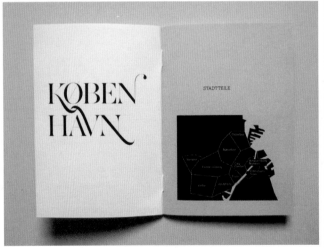

Designer:
**Britta Siegmund**
Nationality:
**Germany**

设计师:
布利塔·齐格蒙特
国家:
德国

Kopenhagen City Guide is a small booklet with insider tips for an exciting stay in Denmark's capital. It contains among others suggestions for nightlife, shopping and sightseeing. It is alone colour print on different coloured paper and packed waterproof on a small plastic envelope.

《哥本哈根城市向导》是一本导游手册,内附以介绍丹麦首都哥本哈根的小贴士,涉及夜生活、购物和观光的相关内容。该手册将一种颜色印刷在不同颜色的纸张之上,运用防水材料进行包装,与精致的塑料封面交相辉映、相得益彰。

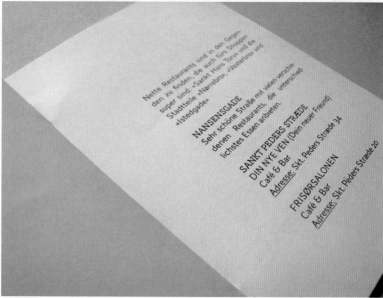

# Nachtvlucht

"夜航" – 城市文化指南手册

Design Agency:
**Nachtbrakers + Elke Broothaers**
Client:
**de Rand**
Nationality:
**Belgium**

设计机构:
Nachtbrakers + Elke Broothaers设计工作室
客户:
德兰特公司
国家:
比利时

Nachtbrakers is a guide to festivities in cultural centres from all around Brussels. You have 6 different kind of disciplines: dance, theatre, festival, senior, classic. For each discipline Elke Broothaers developped a personal colour and pattern, which suits the discipline. These colours and patterns are your guide through the brochure.

"夜航" – 城市文化指南"手册以向读者介绍布鲁塞尔所有文化中心的文化盛宴为主旨。读者在该指南中将欣赏到六种不同的文化主题，包括舞蹈、戏剧、纪念活动、资深人士讲座以及大艺术家讲坛。在每个主题之中，设计师特别开发了一个独特的色调和图案，并使之与主题完美融合。这些贯穿在手册中的色调和图案扮演了向导的角色，引领读者从一个主题过渡到另一个主题。

# New Prevention Technologies (NPT) Toolkit

"全新防御技术(NPT)工具包"手册

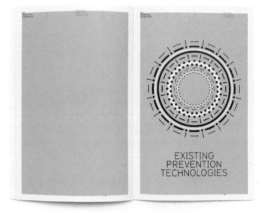

Design Agency:
**COOEE - Leon Dijkstra**
Creative Director:
**Leon Dijkstra**
Designer:
**Leon Dijkstra**
Client:
**GNP+ / NPT**
Nationality:
**The Netherlands**

设计机构:
COOEE －利昂·迪杰斯特拉设计工作室
创意总监:
利昂·迪杰斯特拉
设计师:
利昂·迪杰斯特拉
客户:
GNP+ / NPT组织
国家:
荷兰

New Prevention Technologies (NPT) is a division of GNP+ that supports research and development for people living with HIV. Every prevention technology described has been visually translated into a unique stroke style as part of the identity. In the booklet each chapter contains a combined image made out of the unique stroke-styles and summarises visually its content. The image on the cover is a combination of all chapters.

全新防御技术(NPT)是GNP的一个分支机构,旨在为艾滋病群体提供产品的研究和开发工作。在该项目中,每一个防御技术被形象地诠释成一个独特的笔触,同时作为构成识别的一个部分。在手册中,每个章节包括一个由独特笔画构成的组合图片,从视觉的角度对章节内容进行概括。封面上的图像巧妙地将所有章节的图片综合在一起。

# Team Scope 7 Winning Ways

"团队视野设计工作室的七种成功途径"宣传册

Design Agency:
**Team Scope**
Creative Director:
**Mark Burrough**
Designer:
**Christine Kelly**
Photography:
**Mark Burrough**
Nationality:
**Australia**

设计机构:
团队视野设计工作室
创意总监:
马克·伯勒
设计师:
克里斯汀·凯利
摄影:
马克·伯勒
国家:
澳大利亚

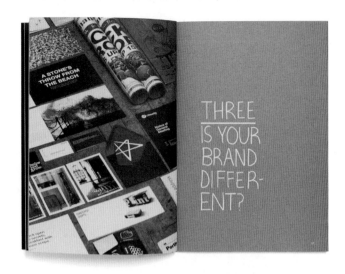

Team Scope celebrated the methodology behind designing and implementing a strong brand. 7 Winning Ways captures the essence of Team Scope's branding ideology – create a clear brand message and identity, dare to be different and create a brand driven culture that is embraced across all areas of your business.

该项目旨在体现团队视野设计工作室在设计和实施一个强有力品牌理念过程中所采用的方法论。该项目牢牢抓住了团队视野设计工作室品牌意识形态的精髓,构建了一个清晰的品牌信息和识别系统,大胆出新、标新立异,创建的品牌驱动文化涉及商业的所有领域。

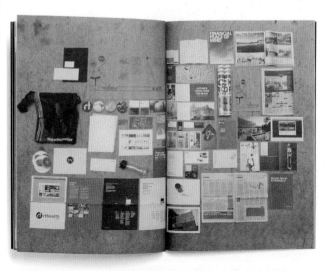

# Che cosa sai sugli inibitori? "你对抑制因子有多少了解？"宣传册

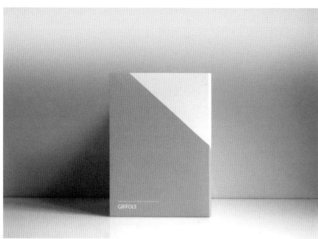

Design Agency:
**CreativeAffairs**
Creative Director:
**Laia Guarro**
Designer:
**CreativeAffairs team**
Client:
**Grífols International SA**
Photography:
**CreativeAffairs team**
Nationality:
**Spain**

设计机构：
CreativeAffairs设计工作室
创意总监：
拉娅·瓜罗
设计师：
CreativeAffairs设计工作室
客户：
Grífols国际组织
摄影：
CreativeAffairs设计团队
国家：
西班牙

Design works for Grifols International dedicated to children who suffer from hemophilia. The commission was to design a booklet that had a certain feature of game to explain the details and possible treatments of this disease. The idea was to use origami as a graphical language to enhance interaction.

该项目专为Grífols国际组织而设计，该组织致力于对血友病患儿童的关注。设计的任务是设计一个广告手册，并使其具有一定的游戏功能，详尽的讲解该疾病的可能性治疗方法。设计的理念是运用折纸工艺作为一种平面语言以强化与读者之间的沟通。

# The Classification of Typefaces: Thermal Study

"字体的分类: 热量研究"宣传册

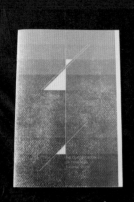

Type specimen booklet is designed to classify typefaces, not only in terms of their period of development, but also upon a thermal scale. Each diverse typeface gives off a unique "feel" based on its physical, historical and sociological qualities. This scale classifies the most prominent typefaces in these groups, and determines their unique thermal essence.

字体样本小册子设计旨在将字体进行明确的分类，依照各字体的开发周期以及热量范围进行划分和分析。每个不同的字体因不同物理、历史和社会学因素的差异，彰显出独特的"质感"。这种尺度能够有效地从这些字体群中筛分出最突出的字体，并确定它们的独特热量实质。

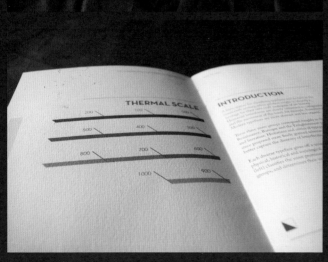

Designer:
**Alex Boland**
Client:
**OCAD University**
Photography:
**Alex Boland**
Nationality:
**Canada**

设计师：
亚历克斯·博兰德
客户：
安大略艺术与设计学院
摄影：
亚历克斯·博兰德
国家：
加拿大

# "Czestochowskie Graffiti"
"琴斯托霍瓦涂鸦设计"宣传册

Design Agency:
**21pixels Studio**
Creative Director:
**Cezary Łopaciński**
Designer:
**Cezary Łopaciński**
Photography:
**Cezary Łopaciński**
Nationality:
**Poland**

设计机构:
21像素设计工作室
创意总监:
赛扎里·罗帕辛斯基
设计师:
赛扎里·罗帕辛斯基
摄影:
赛扎里·罗帕辛斯基
国家:
波兰

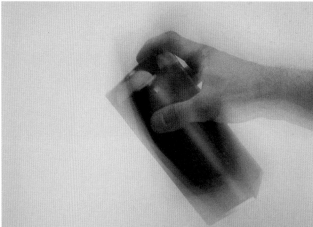

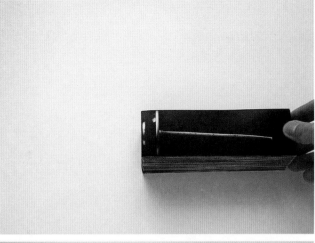

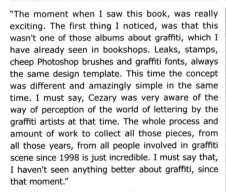

"The moment when I saw this book, was really exciting. The first thing I noticed, was that this wasn't one of those albums about graffiti, which I have already seen in bookshops. Leaks, stamps, cheep Photoshop brushes and graffiti fonts, always the same design template. This time the concept was different and amazingly simple in the same time. I must say, Cezary was very aware of the way of perception of the world of lettering by the graffiti artists at that time. The whole process and amount of work to collect all those pieces, from all those years, from all people involved in graffiti scene since 1998 is just incredible. I must say that, I haven't seen anything better about graffiti, since that moment."

"从我看到这本书的那刻起，我心中就雀跃不已。首先我注意到的是，它与我在书店中所看到的那类涂鸦专辑不同。检查内存泄漏的工具、图章、图像处理软件的画笔以及涂鸦字体等等，常常打造出相似的设计模板。这一次，设计理念别出心裁，真正实现了标新立异，同时其简约的方式更加令人惊叹。我必须承认，赛扎里·罗帕辛斯基对涂鸦设计师的文字世界的感知方式了解得十分透彻。整个加工过程和收集的全部作品所涉及的年代和人物以1998年为起点，这一点的确令人难以置信。而我也必须承认，从这一刻起，我再也没有发现比这个更优秀的涂鸦设计作品。"

# Travel Guides 旅游指南手册

Design Agency:
**Mapas**
Creative Director:
**Daniel Castrejón**
Designer:
**Daniel Castrejón**
Photography:
**Mapas Archive**
Nationality:
**Mexico**

设计机构:
Mapas设计工作室
创意总监:
丹尼尔·卡斯特莱恩
设计师:
丹尼尔·卡斯特莱恩
摄影:
Mapas设计工作室
国家:
墨西哥

Travel Guide for the premium customers of Citi Bank Mexico. The idea was to develop a resistent, fancy and modern travel guide, with his packaging to easy transportation, perfect for long journeys. The design work is based in the colour palette of the flags of the countries, ensuring a clean work, minimal and easy to read.

该项目是专为墨西哥花旗银行的保险客户所设计的旅游指南手册。设计的理念是开发一个与众不同、充满梦幻气息以及时尚现代的旅游指南,并确保其包装考究,便于携带,尤其适合长途旅行者。该设计以国家国旗的配色方案为基础,确保打造干练、简约、易识别的风格气息。

# Democracia Viva - Conferences about Democracy and Europe "民主万岁"-民主与欧洲会议宣传手册

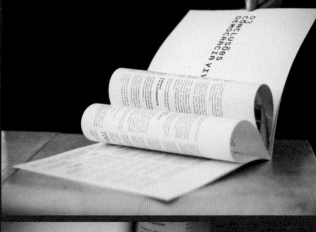

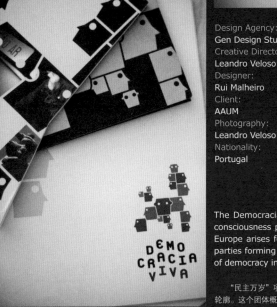

Design Agency: **Gen Design Studio**
Creative Director: **Leandro Veloso**
Designer: **Rui Malheiro**
Client: **AAUM**
Photography: **Leandro Veloso**
Nationality: **Portugal**

设计机构：Gen设计工作室
创意总监：林德罗·维罗索
设计师：瑞·迈克黑罗
客户：AAUM公司
摄影：林德罗·维罗索
国家：葡萄牙

The Democracia Viva project was committed to stimulating the democratic consciousness particularly regarding the European context. The silhouette of Europe arises formed by the different individual perspectives. This notion of parties forming the whole is just as valid for the idea of Europe as to the idea of democracy in which all voices have their weight.

"民主万岁"项目致力于推动欧洲背景下的民主意识。不同的个人见解形成了欧洲大陆的轮廓。这个团体概念对于欧洲的民主理念具有很高的效力，他们拥有绝对的发言权。

# Edinburgh College of Art Prospectus
爱丁堡艺术学院招股宣传册

Design Agency:
**KVGD**
Creative Director:
**Kerr Vernon**
Designer:
**Kerr Vernon**
Client:
**Edinburgh College of Art**
Nationality:
**UK**

设计机构:
KVGD设计工作室
创意总监:
克尔·弗农
设计师:
克尔·弗农
客户:
爱丁堡艺术学院
国家:
英国

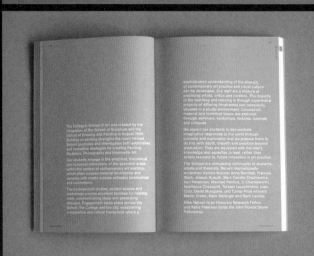

This A5 prospectus design for the Edinburgh College of Art features a blind embossed cover with a black foil down the spine.

该项目专为爱丁堡艺术学院而设计,采用A5格式,百叶窗式压花封面与一个黑色锡箔书脊交相辉映。

# Facebook (let)　"脸谱"作品宣传手册

Design Agency:
Goiaba
Creative Director:
Johannes Fuchs
Designer:
Johannes Fuchs
Photography:
Thuy Duong Phan
Nationality:
Germany

设计机构：
Goiaba设计工作室
创意总监：
约翰·富克斯
设计师：
约翰·富克斯
摄影：
德伊·杜庸·泛恩
国家：
德国

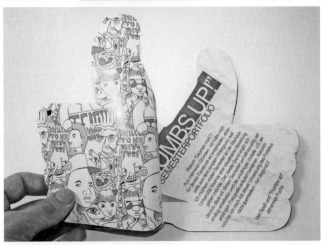

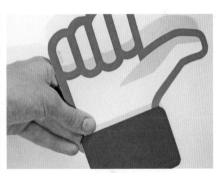

Facebook (let) shows works of Goiaba/Johannes Fuchs that he had created in the time of one semester. He's studying visual communication with the focus on illustration/character design on the Bauhaus-University Weimar/Germany. Why "Thumbs up"? - because the time here in Weimar deserves it!

该项目旨在展现Goiaba设计工作室/约翰·富克斯在一个学期内所创作的作品。目前，他在德国魏玛包豪斯大学插画和人物设计专业学习视觉传达课程。"为什么要设计成翘起大拇指的形态？"因为对于设计师来说，在魏玛学习的这段时间很值得纪念。

# Giants of Typography John Baskerville

"约翰·巴斯卡比尔的大字体"宣传册

Design Agency:
**Billy Blue Student Assignment**
Creative Director:
**Shant Safarian**
Designer:
**Shant Safarian**
Photography:
**istock**
Nationality:
**Australia**

设计机构：
比利·布鲁学生联盟
创意总监：
尚特·萨法里安
设计师：
尚特·萨法里安
摄影：
istock摄影工作室
国家：
澳大利亚

The designer was required to produce a six-page custom-size booklet a little wider than A5 to promote the typeface of a chosen typographer. John Baskerville and his typographic work have always inspired the designer. His work is seen anywhere from modern day movie posters to elegant invitation and books. His work deserves all the praise it receives and more.

该项目旨在打造一个6页的特别尺寸手册，该手册比A5格式略宽，目的在于为一个优秀的字体设计师进行宣传。约翰·巴斯卡比尔及其创作的字体作品为设计师带来了很大的创作灵感。该字体设计师的作品出现在我们生活的很多角落，无论是时尚的电影海报还是优雅的请帖和图书。其作品值得人们去称赞。

# Norrlandsdagarna 2011
Norrlandsdagarna2011年度摄影大会宣传册

Design Agency:
**Mattias Sahlén – Graphic Design & Art Direction**
Creative Director:
**Mattias Sahlén**
Designer:
**Mattias Sahlén**
Client:
**Bergkvist Reklamfoto**
Photography:
**D. Gardner / M. Norman / C. Hedberg / K. Henke / T. Ullberg / N. Jakobsson / S. Harström / J. Kimmich-Javier / M. von Krogh / Klara G/M. Sahlén**
Nationality:
**Sweden**

设计机构:
马蒂亚斯·萨伦 – 平面设计与艺术指导
创意总监:
马蒂亚斯·萨伦
设计师:
马蒂亚斯·萨伦
客户:
Bergkvist Reklamfoto公司
摄影:
D.加德纳, M.诺曼,
C.海德博格, K.汉克,
T.乌尔博格, N.雅各布森,
S.哈斯特罗姆,
J.吉姆米奇·哈维尔,
M.凡·科罗哈, 克拉拉·G,
M.萨伦
国家:
瑞典

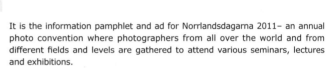

It is the information pamphlet and ad for Norrlandsdagarna 2011– an annual photo convention where photographers from all over the world and from different fields and levels are gathered to attend various seminars, lectures and exhibitions.

该项目是专为Norrlandsdagarna2011年度摄影大会而设计的资料手册和广告。每年，该摄影大会都吸引了全球各地、各行各业的摄影爱好者聚集一堂出席各种研讨会、讲座和展览活动。

# Human / Computer  "人类与电脑"宣传册

Designer:
**Avigail Bahat**
Photography:
**Michael Topyol**
Nationality:
**Israel**

设计师:
艾维盖尔·巴哈特
摄影:
迈克尔·托普伊尔
国家:
以色列

It is the final project in the designer's design studies at Shenkar. A series of posters, a dictionary of associations for terms representing both the computer world and the human world.

该项目是设计师在申卡尔设计与工程学院学习设计课程期间的毕业设计项目。一系列海报、一本介绍相关术语的字典象征着电脑世界和人类世界。

# Seven Things Everyone Should Know About Type

"每个人都应该了解的字体设计的七种规则"宣传册

Design Agency:
**Jo Hoctor**
Creative Director:
**Eamon Spelman**
Designer:
**Jo Hoctor**
Client:
**L.S.A.D**
Photography:
**Jo Hoctor**
Nationality:
**Ireland**

设计机构:
乔·霍克特设计工作室
创意总监:
艾曼·斯佩尔曼
设计师:
乔·霍克特
客户:
利默里克艺术与设计学院
摄影:
乔·霍克特
国家:
爱尔兰

"Typographical design should perform optically what the speaker creates through voice and gesture of his thoughts", EL Lizzitsky said. "Seven things everyone should know about type" is a booklet designed to educate its audience on typographic rules and languages. The step by step guide offers an insight into the practice of typography and its importance within good design. Each rule has a unique icon, which is later presented in the form of a sticker allowing people to interact with the piece and associate each rule with a visual element.

"排版设计应该能够形象地表达出设计者的心声以及思想动态。"

"每个人都应该了解的字体设计的七种规则"是一个广告宣传手册,旨在为读者展现排版设计的规则和语言指南。循序渐进的指导方式引领读者洞悉排版设计的整个过程和方法以及其在一个完美设计项目中所占据的重要地位。每个规则均设计了一个独特的图标,并随后以贴纸的形式拉近读者与项目之间的距离,此外,设计师巧妙地运用了一个视觉元素与每个规则建立起微妙的联系。

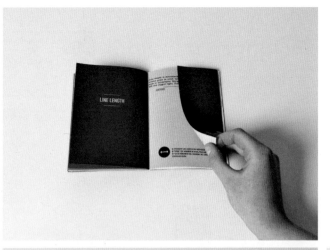

# Kit Pedagógico sobre Género e Juventude

青年性别教学指导手册

Design Agency:
**Gen Design Studio**
Creative Director:
**Leandro Veloso**
Designer:
**Catarina Correia**
Client:
**Conselho Nacional de Juventude**
Photography:
**Leandro Veloso**
Nationality:
**Portugal**

设计机构:
Gen设计工作室
创意总监:
林德罗·维罗索
设计师:
卡塔里那·克雷亚
客户:
全国青年理事会
客户:
林德罗·维罗索
国家:
葡萄牙

The "Kit Pedagógico sobre Género e Juventude" was developed to be used as a tool to promote the gender equality among young people. Considering the user the designers developed a clear layout enriched with illustrations creating an appealing language. To increase usability the pages have different sizes allowing direct access to distinct thematics.

"青年性别教学指导"手册的开发旨在促进年轻群体中男女之间的平等。考虑到受众对象，设计师巧妙地开发了一个清晰的版式结构，并运用丰富的插画创建了一个迷人的视觉语言。为加强该项目的实用性，设计师将页面设计成不同的尺寸，便于读者直接找到鲜明的主题。

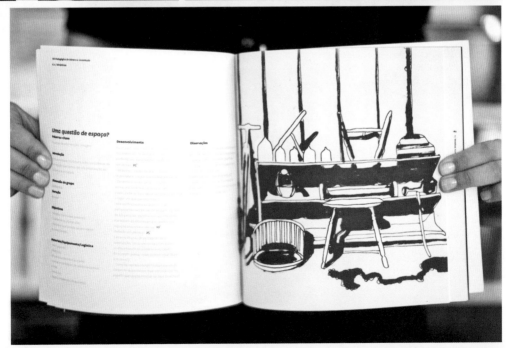

# Manual for Rossmann

Rossmann Polska公司培训手册

Design Agency:
**Studio Baklazan**
Art Director:
**Urszula Kluz-knopek**
Designer:
**Urszula Kluz-knopek**
Nationality:
**Poland**

设计机构:
Baklazan设计工作室
艺术总监:
乌斯左拉·克鲁兹·科诺培克
设计师:
乌斯左拉·克鲁兹·科诺培克
国家:
波兰

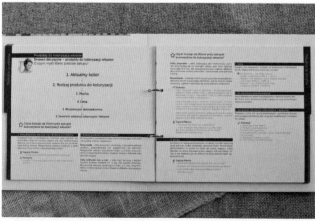

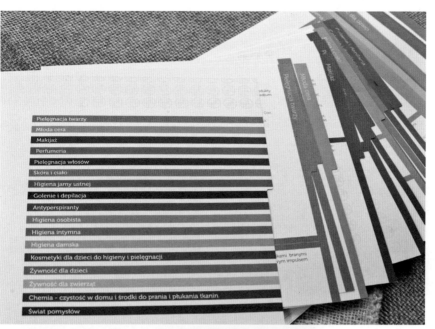

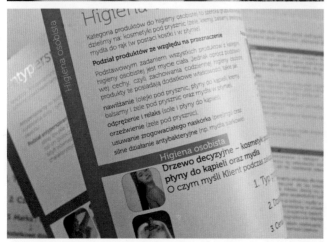

It is an internal training manual designed for Rossman Polska, in the form of a binder, so that those in training can periodically replace the old content with new one. This was directed at a specific professional group, employees of Rossman, who need to be able to have instant access to vital information. On the other hand it is a symbol of a progressive, pragmatic, easily accessible chemist's shop. In order to underline the class and subtlety of Rossmans agenda the designers used lusterless dispersive lacquering (it gives the paper a velvet feel) also each page had its own foiling design.

　　该项目是专为波兰最大的保健与美容连锁店Rossmann Polska公司而设计的内部培训手册，独特的活页夹形式便于培训中新旧内容的周期更换。该手册以独特的专业团体 – Rossmann Polska公司员工为对象，旨在为他们提供快速掌握重要信息的有效途径。另一方面，该项目也象征着一个先进、实用性强、亲切的药店。为突出该公司日常工作事项清晰的分类，设计师运用了亚光发散式涂漆层（使纸张具有天鹅绒般质感），此外，每个页面拥有各自的锡箔质地设计。

# Nuit Blanche 多伦多不眠之夜活动手册

Creative Director:
**Michelle Liando**
Designer:
**Michelle Liando**
Photography:
**Various**
Nationality:
**Canada**

创意总监:
米歇尔·里安多
设计师:
米歇尔·里安多
摄影:
来自各种摄影渠道
国家:
加拿大

A mini booklet that explores and research the success and failures of Toronto Nuit Blanche 2010 based on reviews and interviews during the event. The aim of the report is to find improving points for future Nuit Blanche events.

该项目是一个小型手册,旨在对多伦多不眠之夜2010年度回顾和访谈活动的得失进行探讨和研究。该报告的设计目标是为以后的不眠之夜活动的举办发现改良之处。

# Thesis Proposal  "论文提案" 手册

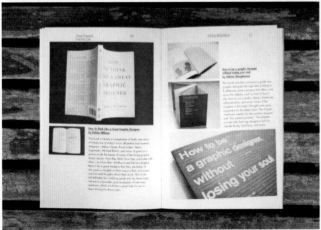

Creative Director:
**Michelle Liando**
Designer:
**Michelle Liando**
Photography:
**Various**
Nationality:
**Canada**

创意总监:
米歇尔·里安多
设计师:
米歇尔·里安多
摄影:
来自各种摄影渠道
国家:
加拿大

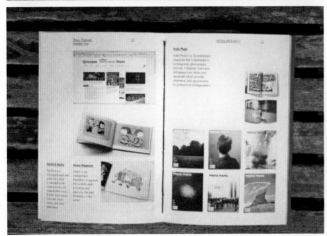

Thesis proposal that is complied into 3 mini booklets, each explores different topics areas. The booklets contain visual and initial research that broadens the knowledge about the thesis topics.

　　该项目包括三个小型手册，每个手册拥有不同的主题。该项目所囊括的视觉和初期研究将论文提案的知识领域进一步拓宽。

# Loipersdorf Booklet 罗尔珀斯多夫温泉手册

Design Agency:
moodley brand identity
Creative Director:
Mike Fuisz / Alex Muralter
Designer:
Sabine Kernbichler
Client:
Thermalquelle Loipersdorf
Photography:
Marco Rossi / Wolfang Croce / Marion Luttenberger
Nationality:
Austria

设计机构：
Moodley品牌识别设计工作室
创意总监：
麦克·弗赛兹、亚历·穆拉尔特
设计师：
萨宾·科恩比其乐
客户：
罗尔珀斯多夫温泉
摄影：
马可·罗西，沃尔夫·克罗齐，马里昂·卢登博格
国家：
奥地利

Such as the general philosophy of the biggest Austrian thermal spa the content of the brochures is also based upon the 3 pillars: "let go", "experience" and "strengthen" – represented in the new CI by the colours blue, orange and lila. Posters, newspaper supplements and accordion-fold booklets were part of a campaign realised by Moodley brand identity leading to a better thermal spa utilisation during the summer months. The look of the brochures is as varied and refreshing as the numerous summer events and leisure facilities offered by the thermal spa.

该项目专为奥地利最大的温泉中心而设计,基于该中心的总体经营理念,该手册的设计同样以三个支柱为基础,即"释放"、"体验"、"强化"。蓝色、橘色和淡紫色,这三种颜色将三种理念鲜明地区分开来。海报、报纸以及折叠式手册均是Moodley品牌识别设计工作室在夏季为罗尔珀斯多夫温泉设计的组成部分。该手册的外观与多种多样的夏日活动以及温泉所提供的休闲设施一样丰富多彩、清爽怡人。

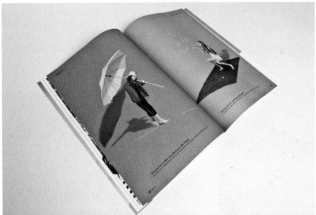

# Número Único "一号"宣传手册

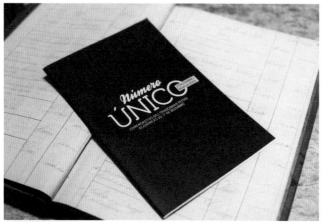

Design Agency:
**Gen Design Studio**
Creative Director:
**Leandro Veloso**
Designer:
**Carla Ribeiro**
Client:
**AAUM**
Photography:
**Pedro Sousa**
Nationality:
**Portugal**

设计机构:
Gen设计工作室
创意总监:
林德罗·维罗索
设计师:
卡拉·里贝罗
客户:
米尼奥大学学生文化表演
摄影:
佩德罗·索萨
国家:
葡萄牙

Every year the 1st of December is celebrated by the Minho University students with a cultural performance. The booklet "Número Único" marks the event and was designed to reflect the tradition in a contemporary way (since it targets young people). This was done by mixing different typography styles in an harmonious way and creating dynamic in the layout composition.

每年的12月1日是米尼奥大学学生文化表演之日。这本小册子以"一号"为名,突出了该活动的举办时间,并运用一个时尚的方式(因为其以年轻一族为导向)彰显出这一活动的传统特色。多样化的字体风格和谐地融为一体,打造了一个充满活力和朝气的版式设计理念。

# Graphic Production Manual 图文制作手册

This is the Art Direction for a Graphic Production book design. The designers use the bright colours to make it more attractive. It's beautiful.

该项目是专为一个平面制作图书设计而提供的艺术指导。设计师运用鲜明的色调使之更具吸引力。该作品的确非常美妙。

Design Agency:
**Loja das Maquetas**
Designer:
**Cátia Alves**
Production Date:
**Conceição Barbosa**
Nationality:
**Portugal**

设计机构：
模型商店设计工作室
设计师：
卡蒂亚·阿尔维斯
客户：
康塞桑·巴博萨
国家：
葡萄牙

# The Big Little Picture

"大图片与小图片"手册

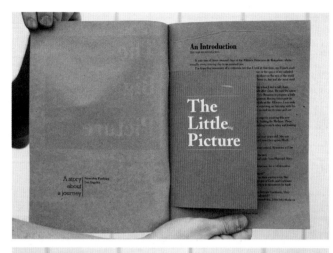
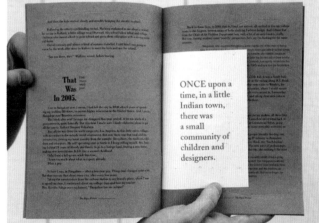

Creative Director:
**Sarmishta Pantham**
Designer:
**Sarmishta Pantham**
Photography:
**Sarmishta Pantham**
Nationality:
**India**

创意总监：
萨梅斯塔·潘达姆
设计师：
萨梅斯塔·潘达姆
摄影：
萨梅斯塔·潘达姆
国家：
印度

The Big Little Picture is one of many steps to come for the designer's proposal towards a teaching system that would explore the possibilities of what he's calling "filtered" input in primary education. Taking the inherent nature of rural education as a model of restraint, economy and sustainability of resources, this system would involve placing deliberate restrictions on ways of learning. It would also encourage children to look to their cultural heritages for inspiration, and so hopefully inspire cultural regeneration through these future citizens. Through the booklet "The Big Little Picture," the designer intend to articulate and present the context and underlying ideas for this schooling system, as a means to invite a discussion with potential collaborators and educators and thereby prepare a case for its further development.

"大图片与小图片"项目是设计师所开发的一个教学系统中的一个环节，这个教学系统旨在证明于基础教育过程中植入所谓的"过滤"模式的可行性。以乡村教育的内在本质作为经济和可持续性资源受限为例，这一系统能够取代学习过程中的人为限制。另外，该系统还能够鼓励孩子们从他们的文化遗产中寻找灵感，从而，让文化在这些世界未来的主宰者手中得以再生。"大图片与小图片"手册巧妙地传达出设计师提出的这一教学系统的环境和潜在理念，并作为一条纽带，将其与潜在的合作伙伴和教学工作者维系在一起，从而促进其进一步的开发。

# Slab-Serif  "雕版衬线字体"宣传册

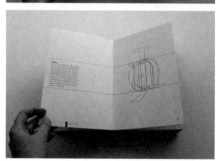

| Creative Director: | 创意总监： |
| --- | --- |
| Caroline Sauter | 卡罗琳·索特 |
| Designer: | 创意总监： |
| Caroline Sauter | 卡罗琳·索特 |
| Photography: | 摄影： |
| Reserche | Reserche组织 |
| Nationality: | 国家： |
| Germany / Swissland | 德国，瑞士 |

It is an A5 Booklet and it's about the history of slab-serif fonts and their development. It is a collection of the most important and influencing fonts.

该项目是一个A5格式的小册子，旨在介绍雕版衬线字体的历史以及发展过程。其中囊括了诸多极为重要以及颇富影响力的字体。

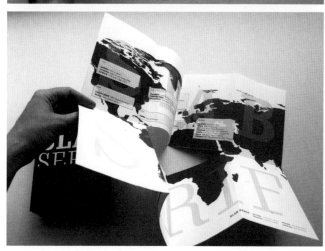

# Children's Books and Illustration from Lithuania

立陶宛儿童图书插画宣传册

**Design Agency:**
PRIM PRIM
**Creative Team:**
Migle Vasiliauskaite,
Kotryna Zilinskiene
**Nationality:**
Lithuania

设计机构:
PRIM PRIM设计公司
创意团队：
米格尔·瓦斯里奥斯凯特、
科特丽娜·泽林斯基尼
国家:
立陶宛

This is the booklet design for Bologna Book Fair 2012. The minimalistic booklet design would accord with the white and clean Lithuania's stand space in Bologna Book Fair and wouldn't compete with the colourful illustrations inside the pages. The publication has reporter style binding, and embossed cover design.

该项目是为2012博洛尼亚书展所设计的图书宣传册。简约的手册设计与立陶宛在书展上的简洁的白色展位相得益彰，与内页的彩色插画形成了鲜明对比。手册采用了竖翻式装订，封面采用凸字设计。

# Abundant Australia  "地大物博的澳洲"宣传册

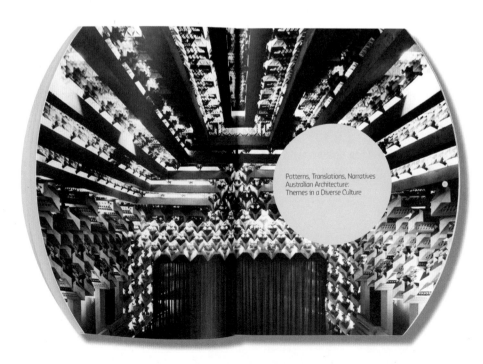

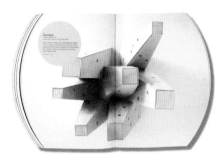

Design Agency:
**Frost design**
Client:
**RAIA - Abundant Australia, Venice Biennale**
Nationality:
**Australia**

设计机构:
**弗罗斯特设计工作室**
客户:
**澳大利亚皇家建筑师学会 – 地大物博的澳洲与威尼斯双年**
国家:
**澳大利亚**

The Creative Directors collaborated on all curatorial elements, making the exhibition as experiential as possible. The exhibition responds to the Biennale director Aaron Betsky's theme: "Out there: architecture beyond building", with the whole Pavilion becoming an installation to be experienced. Graphics applied to the front of the building, "Abundant Australia" repeated on a horizon line, unmistakably announcing arrival at the Australian Pavilion. The designers worked with Dulux to develop a new paint colour named "Abundant", which was applied to the building exterior and interior making a strong branded statement. The vibrant yellow colour provides an instant sense of lively "Australia-ness".

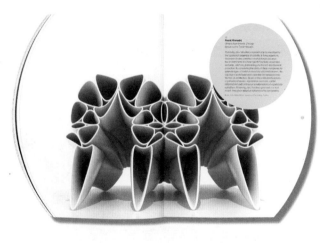

创意总监与所有的管理人物进行合作，尽可能地将展览打造得更为实际。该展览旨在响应双年展主管亚伦·贝特斯奇提出的主题，即"在那里：建筑学超越了建筑本身的意义，而整个展厅变成了一个体验装置"。建筑前端运用的图案 – "地大物博的澳洲"沿着横向重复，鲜明地宣告了澳大利亚展厅所在地。设计师与多乐士涂料公司合作，共同开发了一个新型彩色涂料，并命名为"富足"，并将其应用到建筑的外部和内部，巧妙地打造了一个强有力的品牌宣言。充满活力的黄色色调明快地展现了澳大利亚的生活、活泼气息。

# Anorexia Nervosa "神经性厌食症"宣传册

Design Agency:
**Alistair Stephens Design**
Creative Director:
**Alistair Stephens**
Designer:
**Alistair Stephens**
Client:
**Beat - Beating Eating Disorders**
Photography:
**Alistair Stephens**
Nationality:
**UK**

设计机构:
阿利斯泰尔·斯蒂芬设计工作室
创意总监:
阿利斯泰尔·斯蒂芬
设计师:
阿利斯泰尔·斯蒂芬
客户:
击败—击败饮食失调慈善机构
摄影:
阿利斯泰尔·斯蒂芬
国家:
英国

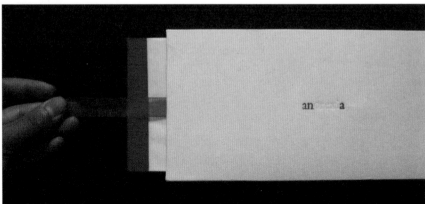

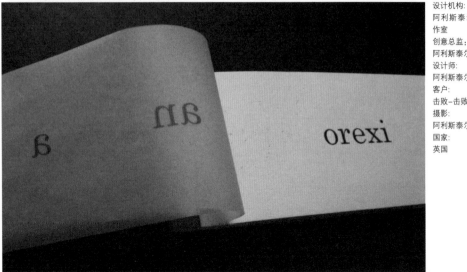

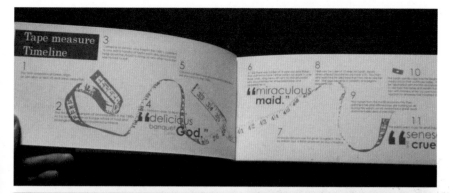

In the wake of many pro-anorexia web sites, the designer investigated into the causes and effects of anorexia. Playing with typography and shocking visuals, in an engaging book format, supported by the Beat Charity – beating eating disorders.

仿效很多领先的厌食症治疗网站，设计师对该疾病的起因和不利影响进行了深入调查。该项目巧妙运用灵活的字体、醒目的视觉元素以及充满朝气的书籍版式，十分引人注目，该设计获得了"击败–击败饮食失调"慈善机构的鼎力推荐。

# Zlatko Prica Monograph "兹拉特科·普里察专著"宣传册

Design Agency:
**Sensus Design Factory Zagreb**
Creative Director:
**Nedjeljko Spoljar**
Designer:
**Nedjeljko Spoljar / Kristina Spoljar**
Client:
**Kabinet grafike HAZU**
Nationality:
**Croatia**

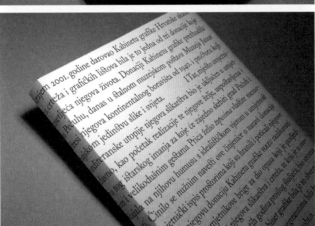

设计机构:
萨格勒布Sensus设计工厂
创意总监:
耐蒂艾尔克·斯颇里扎
设计师:
耐蒂艾尔克·斯颇里扎,克莉斯汀娜·斯颇里扎
客户:
Kabinet grafike HAZU博物馆
国家:
克罗地亚

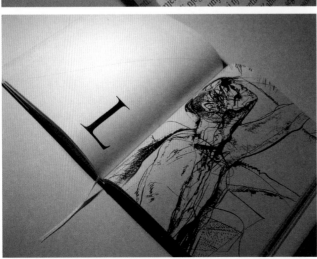

The monograph presents selected works from the author's donation to the Department of Prints and Drawings at the Croatian Academy of Sciences and Arts. Representation of various graphic techniques caused the division into sections, each with a separate page and characteristic illustration. Special attention is devoted to defining details, text layout and treatment of typography, especially in places where it is often neglected, such as footnotes and the signatures next to works, biographies... The cover contains a block of text from the introduction and gives direct information about the content, avoiding the conventional principles of cover treatment. The title is incorporated in the introduction text and stands out only by its colour, while the name of the author of the book is located at the back of the book.

该专题著作向读者展现了捐赠给克罗地亚科学与艺术学院印刷和绘画专业的精选作品。不同的平面设计技巧将该著作划分成若干部分，每个部分拥有一个独立的页面和人物插画。关键细节、文字排版、字体的处理等被赋予了极大的关注，尤其是容易忽略的地方，例如脚注、著作旁边的签名、传记等等。封面上援引了内容简介的部分文字，直接将内容信息呈现给读者，打破了传统的封面处理模式。标题与文字简介融为一体，以不同的色调突出所在位置，而该书的作者名字则被设置在了书后。

# NESTA

"NESTA教学用书"宣传册

"Launch your own successful creative business" is an educational resource containing four booklets, worksheets and tutor notes. The booklets and posters were printed using fluro pantones and a black. The casing was white foil printed onto heavyweight dutch grey board.

"开发自己的成功创意事业"是一个教学用书,其中包括4个小册子,工作笔记和辅导记录。手册和海报均运用了荧光潘通色卡和黑色调印刷。书套采用白色烫印处理,以厚重的荷兰灰纸板为原料。

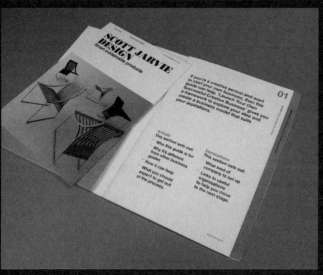

Design Agency:
**Kerr Vernon Graphic Design**
Designer:
**Kerr Vernon**
Client:
**NESTA**
Nationality:
**UK**

设计机构:
克尔弗农平面设计工作室
设计师:
克尔弗农
客户:
NESTA教学
国家:
英国

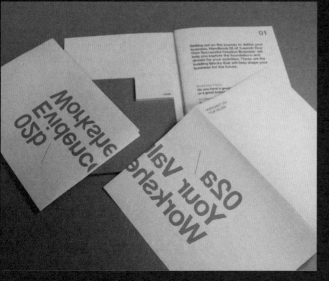

# Alternative Project   "备选项目"手册

**Design Agency:**
Shenkar - College of Engineering & Design
**CreativeDirector:**
Yonatan Solomon
**Designer:**
Yonatan Solomon
**Nationality:**
Israel

设计机构:
申卡尔工程与设计学院
创意总监:
约拿坦·所罗门
设计师:
约拿坦·所罗门
国家:
以色列

A set of 3 booklets depicting biographies of 3 characters that represent the creative and innovative spirit of their era. The design of each booklet refers to its character life developments and provides experimental reading.

该项目包括三个广告宣传手册,以彰显三位能够代表他们所处时代的创意和创新精神的杰出人物传记为主题。每个手册的设计以展现每个人物性格的历练和进步过程为基础,旨在为读者提供最佳的阅读分析体验。

# Korte 50th Anniversary Book

"科特公司成立50周年图书设计"宣传册

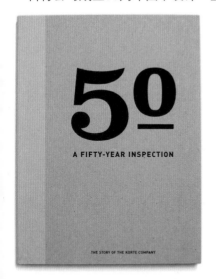

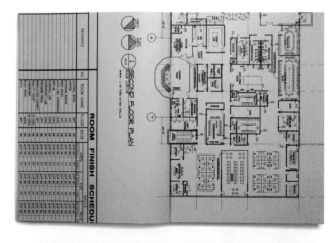

Design Agency:
**Knoed Creative**
Art Director:
**Kim Knoll**
Designer:
**Kim Knoll**
Client:
**The Korte Company**
Nationality:
**USA**

设计机构:
Knoed创意设计工作室
艺术总监:
吉姆·科诺尔
设计师:
吉姆·科诺尔
客户:
科特公司
国家:
美国

The Korte Company is a design-build construction company with more than 50 years of experience. In celebration of their golden year, this booklet was designed to showcase the highlights and hardships from 1958-2008. Each page contains elements that continue from one page to the next, creating one long continuous timeline. Other collateral designed to mark the occasion included an anniversary logo, patch, stationery, notepad, sketch pad, print/banner ads and custom temporary tattoos.

  Korte公司是一个建筑设计公司,拥有50多年的创办历史。为庆祝该公司黄金50周年的到来,该手册的设计旨在彰显该公司自1958年至2008年间所经历的喜悦和艰辛。每个页面上所蕴含的元素从一个页面上延伸至下一页,巧妙地构建出一个连续的时间表。其他附属资料的设计以突出这一重大时刻为基础,包括周年纪念标志、补丁、文具、记事本、速写本、印刷横幅广告和特别设计的临时花纹等。

# Booklet of the School of Art Saint Luc

圣吕克艺术学院手册

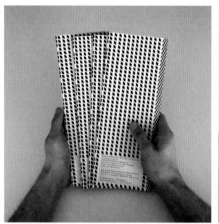

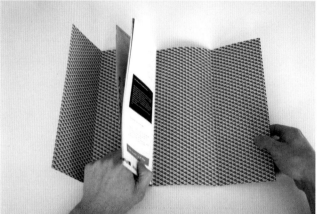

Design Agency:
**BENBENWORLD**
Designer:
**Benot Bodhuin**
Client:
**ESA Saint Luc**
Nationality:
**France**

设计机构：
BENBENWORLD设计工作室
设计师：
贝诺特·博得胡恩
客户：
圣吕克艺术学院
国家：
法国

It is the design of graphic identity of the art college Saint Luc of Tournai. Creation includes logo, pictograms, lettering, patterns, typography ... language applied in particular to the booklet and the website of the school.

该项目是专为图尔奈圣吕克艺术学院而设计的平面识别系统。项目涉及标识、图标、字体、图案、印刷品等方案设计，独特的设计语言专为学院的手册和网站量身打造。

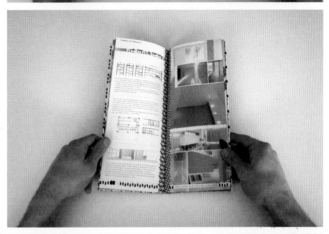

# Klavir.info Portfolio

Klavir.info作品型录

Design Agency:
**Kolektiv Studio**
Designer:
**Lukas Kijonka / Michal Krul**
Client:
**Piano33, www.klavir.info**
Nationality:
**Czech**

设计机构:
Kolektiv设计工作室
设计师:
卢卡斯·基约卡，米甲·克鲁尔
客户:
Piano33团队, www.klavir.info
国家:
捷克

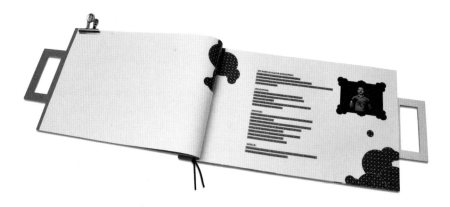

This is the catalog work of industrial designer Mark Kedziersky. The client requested to add more work into the presentation and therefore it was designed to be an easily separable form with cutouts. The lines used elastic string.

该项目是专为工业设计师马克·凯德兹尔基而设计的型录。客户要求设计能够在描述过程中添加更多的作品,因此,设计师巧妙运用剪贴画打造一个易拆分的形态。线条运用橡皮筋而设计。

# New G(o)ods "新设计作品"型录

Design Agency:
**Kolektiv Studio**
Designer:
**Michal Krul / Jan Kosátko / Lukas Kijonka**
Client:
**Designcabinet**
Nationality:
**Czech**

设计机构:
**Kolektiv设计工作室**
设计师:
**米甲·克鲁尔，简·库萨蒂科，卢卡斯·基约卡**
客户:
**设计陈列室**
国家:
**捷克**

It is Product Contest Student National Award for Design, organised by Design Cabinet CR.
It includes awards nominated in the category Graphic Designer of the Year.

该项目专为"设计陈列室"举办的国内学生设计大赛而设计。该项目中囊括了本年度平面设计领域的所有获奖提名作品。

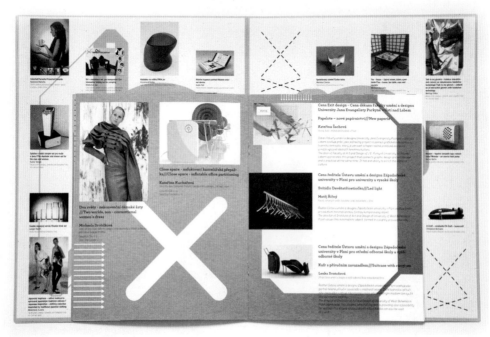

# 15 Años de Fotografía Española Contemporánea

第十五届西班牙当代摄影展型录

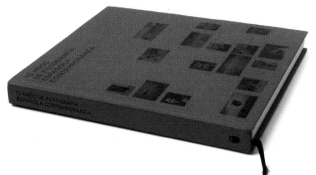

Design Agency:
**Erretres**
Designer:
**Erretres**
Client:
**City of Alcobendas**
Nationality:
**Spain**

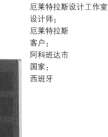

设计机构:
厄莱特拉斯设计工作室
设计师:
厄莱特拉斯
客户:
阿科班达市
国家:
西班牙

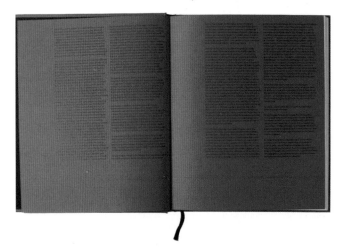

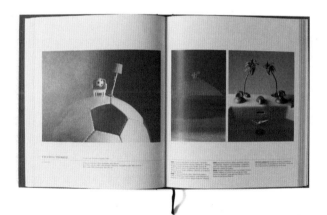

Erretres was in charge of designing the catalogue for the exhibition, 15 Años de fotografía contemporánea (15 years of Spanish contemporary photography) showcasing the City of Alcobendas' collection of photographs from Spain's contemporary photographers.

设计师厄莱特拉斯受邀为第十五届西班牙当代摄影展提供型录设计，以展现阿科班达市从西班牙当代摄影师手中所搜集到的全部摄影作品为主旨。

# Solaris Look Book　Solaris公司艺术画册型录

| | |
|---|---|
| Design Agency: | 设计机构: |
| Etolle Rouge | Etoile Rouge设计工作室 |
| Creative Director: | 创意总监: |
| Roni Bitran | 罗尼·彼特兰 |
| Designer: | 设计师: |
| Roni Bitran | 罗尼·彼特兰 |
| Client: | 客户: |
| Solaris | Solaris公司 |
| Photography: | 摄影: |
| Philippe Sebirot | 菲利普·赛比洛特 |
| Nationality: | 国家: |
| France | 法国 |

It is the art direction for Solaris, sunglasses brand (look book). The design work includes types treatment, graphic layout and art direction for the shoot.

该项目是专为Solaris公司——一个太阳镜品牌而设计的艺术指导。字体的设计、平面版式以及艺术指导均以摄影为导向。

# Catalogue Fall Winter 秋冬型录

Creative Director:
**Carlos Ribeiro**
Designer:
**Carlos Ribeiro**
Client:
**Salsajeans**
Photography:
**Sérgio Matos**
Nationality:
**Portugal**

创意总监：
卡洛斯·里贝罗
设计师：
卡洛斯·里贝罗
客户：
Salsajeans公司
摄影：
塞尔吉奥·马托斯
国家：
葡萄牙

Catalogue presentation of collection Fall Winter. This catalogue marked a new stage in Salsajeans, its rebranding. In this catalogue were passed the new values of the brand and the colour (grey and red). This catalogue was produced in fine papers with double sheet printed on the inside.

　　该项目是专为Salsajeans公司秋冬季服饰系列而设计的型录。该型录标志着Salsajeans公司在品牌重塑之后上升到一个崭新的高度。该型录传达了这一品牌的全新价值理念，并突出服饰的主打色调（灰色和红色）。这一目录以优良的纸张为原料，内页采用双面印刷。

# Central Periphery Exhibition Catalogue
中央边缘展览会型录

Design Agency:
**Trampoline Design Pty Ltd.**
Creative Director:
**Sean Hogan**
Designer:
**Sean Hogan**
Client:
**RMIT Gallery, RMIT University**
Photography:
**Mark Ashkanasy**
Nationality:
**Australia**

设计机构:
蹦床设计有限公司
创意总监:
肖恩·霍根
设计师:
肖恩·霍根
客户:
墨尔本皇家理工大学画廊
摄影:
马克·阿什坎纳西
国家:
澳大利亚

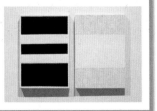

Trampoline designed the catalogue for the exhibition Central Periphery for RMIT Gallery. The catalogue is a 200mm x 200mm perfectly bound booklet of 48 pages, and is designed using bold and minimalist typography to complement the minimalist aesthetic of the artwork. Typeface used throughout was Akzidenz Grotesk.

该项目是蹦床设计有限公司专为墨尔本皇家理工大学画廊举办的中央边缘展览而设计的型录。该目录的尺寸为200×200毫米，装订完美，共计48页，运用粗体和抽象主义字体巧妙地为艺术品营造出简约的美感。整个设计中所采用的是Akzidenz Grotesk字体。

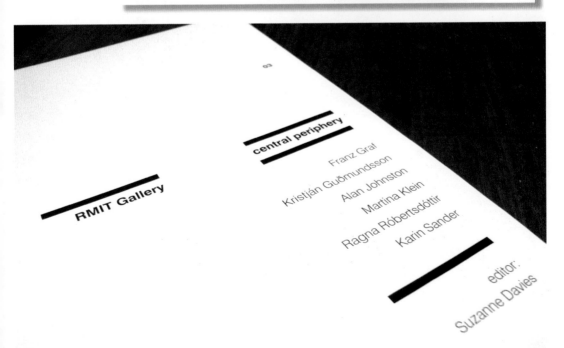

# Italbox 2011

Italbox 公司2011型录

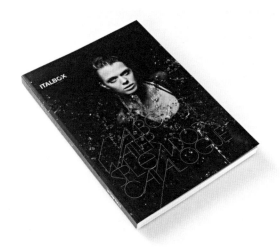
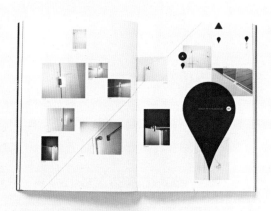
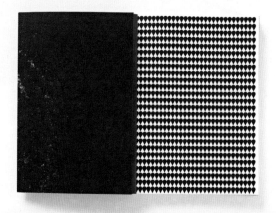

Design Agency:
**Forma**
Creative Director:
**Nuno Cunha**
Designer:
**Nuno Cunha**
Client:
**Italbox, Viriato & Viriato**
Photography:
**Viriato & Viriato**
Nationality:
**Portugal**

设计机构：
Forma设计工作室
创意总监：
努诺·库尼亚
设计师：
努诺·库尼亚
客户：
Italbox公司，
维里亚托＆维里亚托设计公司
摄影：
维里亚托＆维里亚托设计公司
国家：
葡萄牙

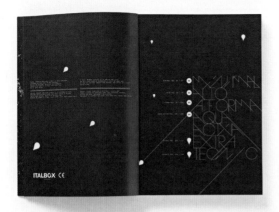

There is no dissociation in the possible use of water as a crosscutting theme in the communication of the spirit underlying the concept of the product - Italbox - the water protect, water and life. The fluidity of life and the spaces that focus. The intimate space afforded by the bath shower and the positive energy, the temptation to control in a pleasant, comfortable, safe, beautiful… The bath and the shower, are moments of intimacy and happiness daily. The body clean, fresh are energised by the substance of life- water.

该项目的设计理念以水为主题，突出了Italbox公司产品的精髓所在，即强调对水源的保护，关注水源即是对生命的重视。整个设计以流畅的生活和空间为重点。柔和的淋浴和正能量使深处在私密空间中的人们深刻感受愉悦、舒适、安全、美妙……享受沐浴的时刻是一天中最惬意、快乐的时刻之一。身体的洁净、清爽全部要归功于生命物质——水体的存在。

# Getxophoto

Getxophoto国际摄影节型录

Design Agency:
**Barfutura**
Designer:
**Sergio Gonzalez Kuhn**
Client:
**Getxophoto**
Nationality:
**Spain**

设计机构:
Barfutura设计工作室
设计师:
塞尔吉奥·冈萨雷斯库恩
客户:
Getxophoto摄影节
国家:
西班牙

It is the Getxophoto. Catalogue and poster for the International Photo Festival.

该项目是专为Getxophoto国际摄影节而设计的型录和海报。

# Mujeres en Plural    "女性群体"型录

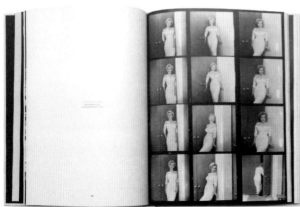
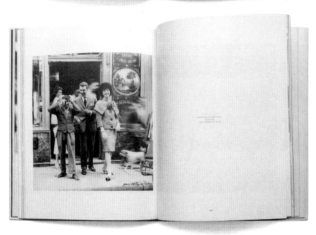

Design Agency:
**Erretres**
Designer:
**Erretres**
Client:
**Fundación Canal Isabel II**
Nationality:
**Spain**

设计机构:
厄莱特拉斯设计工作室
设计师:
厄莱特拉斯
客户:
伊莎贝尔二世运河基金会
国家:
西班牙

The Lola Garrido Collection funded Mujeres en plural, an exhibition highlighting the works of famous photographers like Robert Frank and Irving Penn, which portrayed women through time. The designers designed and oversaw the production of the catalogue, its graphic image as well as the environmental graphics where the exhibition took place.

由萝拉·加里多收藏馆投资举办的女性群体展以展现罗伯特·弗兰克和欧文·佩恩等著名摄影师的作品为主题，这些摄影师一直致力于关注女性的生活和现状。设计师受邀为该展览提供型录、平面形象以及展览举办地的环境平面设计方案。

# Contemporary Craft from Lithuania

立陶宛当代手工设计型录

Design Agency:
**PRIM PRIM**
Creative Team:
**Migle Vasiliauskaite, Kotryna Zilinskiene**
Nationality:
**Lithuania**

设计机构：
PRIM PRIM设计公司
创意团队：
米格尔·瓦斯里奥斯凯特、科特丽娜·泽林斯基尼
国家：
立陶宛

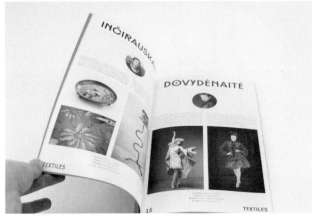

This is a catalog and tote bag design presenting Lithuanian artists participating at the Craft Show, Philadelphia Museum of Art, USA. Colour of the catalog is Lithuanian mint green. The main design element – typeface is used for the headers. The note "Art by Hand" embossed on the cover for the pleasant touch at your fingertips.

该项目是为立陶宛艺术家参加美国费城美术馆所举办的手工设计展所设计的型录和手提袋设计。型录的色彩选用了立陶宛具有代表性的薄荷绿。各个标题都采用了独特的字体。注解"手工艺术"几个字从封面上凸起，提供了独特的触感。

# Tom Dixon Catalogue 汤姆·迪克逊公司产品型录

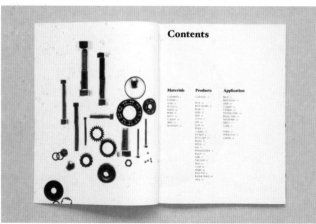

Design Agency:
Mind Design
Creative Director:
Holger Jacobs
Designer:
Johannes Höhmann
Client:
Tom Dixon
Photography:
Henry Bourne
Nationality:
UK

设计机构:
思想设计工作室
创意总监:
霍尔格·雅各布斯
设计师:
约翰·霍曼
客户:
汤姆·迪克逊
摄影:
亨利·伯恩
国家:
英国

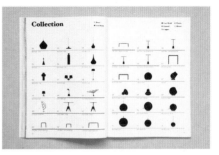

It is Tom Dixon collection catalogue. Most furniture designers organise their collections by product groups (chairs, tables, lamps, etc.) but this catalogue is divided into material groups (steel, copper, brass, etc.) as Tom Dixon often produces different items with the same industrial production method.

The catalogue features large illustrated initial letters showing the various production processes and materials behind the products. The illustrations were inspired by the work of the little-known German designer Max Bittrof in the 1920s.

该项目是专为汤姆·迪克逊公司产品设计的型录。很多家居设计师喜欢将生产的产品按照各自的功能进行区分（例如椅子、桌子、灯具等），而该目录中则依据它们的制成材料进行划分（例如钢、铜、黄铜等），而这恰恰与汤姆·迪克逊公司运用相同的工业加工方式生产不同的物品相得益彰。

该型录运用了大幅插图大写字母以展现多样化的加工过程和产品背后的原料。插画的设计受到了20世纪20年代鲜为人知的德国设计师马克斯·贝特洛夫设计的作品的启发。

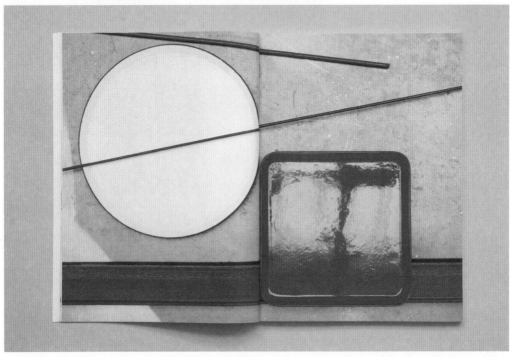

# Dace Lookbook Fall 戴斯品牌秋季型录

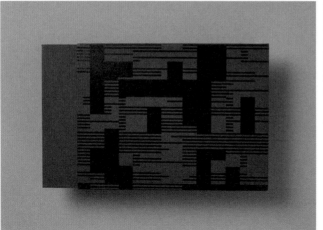

Design Agency:
**Xavier Encinas Studio**
Creative Director:
**Xavier Encinas**
Client:
**Dace**
Nationality:
**Canada**

设计机构:
泽维尔·恩西纳斯设计工作室
创意总监:
泽维尔·恩西纳斯
客户:
戴斯品牌
国家:
加拿大

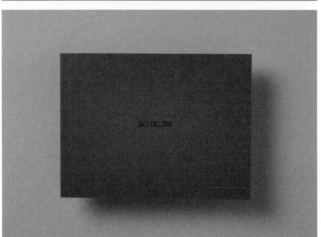

It is the Art Direction & Graphic Design for Dace, women's fashion label.

该项目是专为戴斯女性时装品牌而设计的艺术指导和平面设计方案。

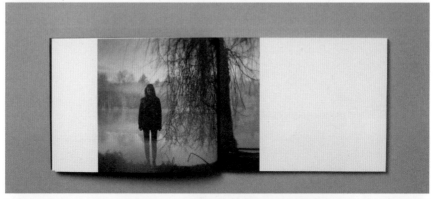

# De Ideale Man  "完美男人"型录

Designer:
**Roosje Klap**
Client:
**Gemeentemuseum Den Haag**
Nationality:
**The Netherlands**

设计师:
卢赛·克莱普设计工作室
客户:
海牙市
国家:
荷兰

Choose your own Ideal Man with one of the four interchangeable covers wrapped around this book. Inside you will find contagious fashion for men from the 17th century up to today: from Galliano to Bernhard Willhelm and from Yves Saint Laurent to "sporty" fashion. The exhibition was designed by Maarten Spruyt, with whom Roosje Klap also did the art direction for of Verschurens' cool photographs. Roosje Klap was also responsible for the routing and the typography in the exhibition.

"从四个可更换的图书封面中选择自己最喜欢的理想男人类型"。从这本书中读者将会欣赏到17世纪至今男装的变迁:从加利亚诺到本哈德·威荷姆,从伊夫·圣·洛朗服饰再到运动装。该展览由玛坦·斯普鲁特和卢赛·克莱普设计工作室共同设计,此外,卢赛·克莱普设计工作室还为摄影师Oof Verschurens的超酷图片提供艺术指导。另外,该工作室还为展览提供了路线和字体设计方案。

# Introspection 2010-2012. Lithuanian Architecture Exhibition

"2010-2012立陶宛建筑展——反省"型录

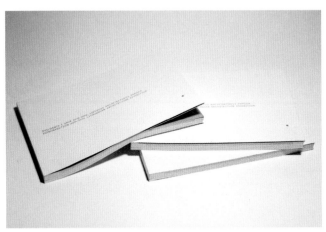

Design Agency:
**PRIM PRIM**
Creative Team:
**Migle Vasiliauskaite, Kotryna Zilinskiene**
Nationality:
**Lithuania**

设计机构:
PRIM PRIM设计公司
创意团队:
米格尔·瓦斯里奥斯凯特、科特丽娜·泽林斯基尼
国家:
立陶宛

This the catalog design for an exhibition "Introspection 2010-2012. Lithuanian Architecture Exhibition". Design concept connects with the symbolic prize of the exhibition – a so called sculpture "the Meter" which is given to the 5 winners of the competition. In the visuals of the website and catalog the designers play with letter "I" which reminds the silhouette of this award.

该项目是为"2010-2012立陶宛建筑展——反省"所设计的型录。设计概念与展览所编颁发的设计大奖结合起来。设计师在展览的网站和型录设计中巧妙地融入了字母"I",与奖杯的剪影十分相似。

# KTM Catalogue KTM型录

Design Agency:
**Tamer Köşeli**
Client:
**Kale Design Centre**
Photography:
**Tamer Koseli**
Nationality:
**Turkey**

设计机构:
泰默·科赛利设计工作室
客户:
凯勒设计中心
摄影:
泰默·科赛利
国家:
土耳其

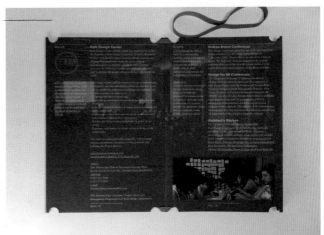

Catalogue contains information about Kale Design Centre in two language—Turkish, English. One side is Turkish the other is English.
The main purpose of design is when you reading information in language you want you couldn't disturb by other language. You can read side you want by removing rubber binding.

该型录中用土耳其语和英语两种语言对凯勒设计中心进行介绍。一面是土耳其语，一面是英语。
设计的主要目的是当读者在阅读一种语言的信息过程中不会被另一种语言所打扰。读者可以移动橡皮筋选择任何一侧进行阅读。

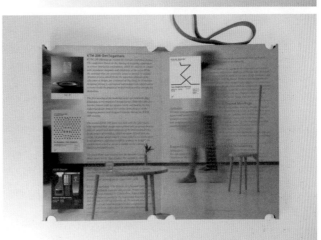

# Catalogue 2011 2011年型录

Designer:
**Danielle Blay / Kelly Satchell / Ally Carter / Isaac Konczak / Elliott Denny**
Photography:
**Danielle Blay / Kelly Satchell / Ally Carter / Isaac Konczak / Elliott Denny**
Nationality:
UK

设计师:
丹妮拉·布雷,凯莉·萨特彻尔,艾莉·卡特,以撒·肯扎克,艾略特·丹尼
摄影:
丹妮拉·布雷,凯莉·萨特彻尔,艾莉·卡特,以撒·肯扎克,艾略特·丹尼
国家:
英国

It is the University of Brighton Graphic Design & Illustration Graduate Catalogue 2011, and also a printed showcase of the work produced and exhibited by university graduates from the Graphic Design and Illustration BA hons courses. Each double page spread displays specially selected work from each individual graduate's exhibited work. The publication was designed to capture and reflect the varied and eclectic body of work produced by Brighton graduates.

该项目是专为布莱顿大学平面设计和插画设计专业毕业生而设计的2011年目录。目录中展现了平面设计和插画设计专业学士毕业生的设计作品。每个跨页特别介绍了每个毕业生的精选作品。该出版物的设计旨在捕捉和彰显布莱顿大学毕业生多样化和折衷主义的作品风格。

can containing angry wasp.

# Ekologgruppen (The Ecologist Group)

Ekologgruppen(生态学家集团)型录

Design Agency:
**Kollor**
Art Director:
**Erik Tencer**
Designer:
**Åse Ekström**
Photography:
**Petter Ericsson**
Nationality:
**Sweden**

设计机构:
Kollor设计工作室
艺术总监:
艾瑞克·坦赛
设计师:
埃塞·埃克斯特朗
摄影:
皮特·爱立森
国家:
瑞典

Ekologgruppen works with assignments within environment and nature preservation. This project, the river of Kävlinge, is cooperation between nine communities in southern Sweden, all situated in the delta of the river of Kävlinge. To design a publication for this project, Ekologgruppen turned to Kollor. The ambition was to make the content more readily available by using an exciting format, more ways-in and a comprehensive pictorial material. The publication is printed on environmentally compensated paper and was distributed within the project, to organisations for the preservation of nature and environment and to county and governmental administrations.

Ekologgruppen集团致力于环境和自然的保护工作。
该项目以介绍Kavlinge河为主题，是位于瑞典南部Kavlinge河三角洲地区九个团体共同努力的结晶。随后，Ekologgruppen集团委托Kollor设计工作室为其提供出版物的设计。设计的主要目标是运用一个充满活力、时尚的版式，创建更为易懂的文章内容，并运用综合的绘画材料。该出版物以环保纸张为原料，并在项目、县级和政府行政机构中分发，组织和宣传对自然与环境的保护。

# Hemostasis Product Catalogue 止血产品型录

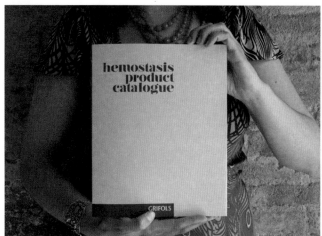

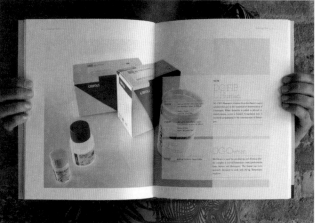

Design Agency:
**CreativeAffairs**
Creative Director:
**Laia Guarro**
Designer:
**CreativeAffairs team**
Client:
**Diagnòstic Grífols SA**
Photography:
**CreativeAffairs team**
Nationality:
**Spain**

设计机构:
CreativeAffairs设计工作室
创意总监:
拉娅·瓜罗
设计师:
CreativeAffairs创意团队
客户:
Diagnòstic Grífols SA学院
摄影:
CreativeAffairs创意团队
国家:
西班牙

Diagnostic Grifols SA commissioned CreativeAffairs the general product catalogue for the Hemostasis division and to be in charge of the arts direction and graphic design of it.
The brief was to create a new graphical language that stepped out from the stereotypes that are usually represented in the pharmaceutical world. Therefore, the design aim was to make it look like an architectural catalogue or a design book rather than just a technical document.

　　Diagnòstic Grífols SA学院委托CreativeAffairs设计工作室为其提供止血部综合产品名录,并负责该设计的艺术指导和平面设计工作。
　　设计的理念是创建一个全新的平面设计语言,打破原有制药行业的传统模式。因此,整个设计如同一个建筑目录又或者是一本设计类图书,而不仅仅是一个教学文献。

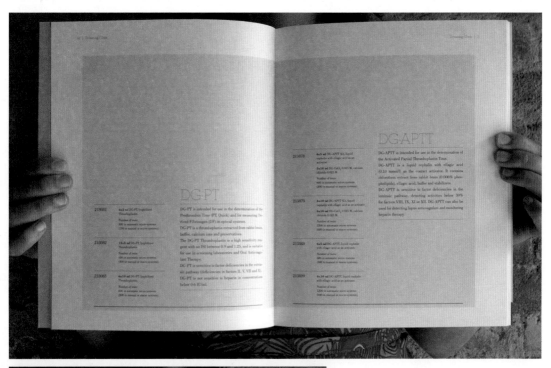

# Lighten Style Product Brochure 特色照明产品宣传册

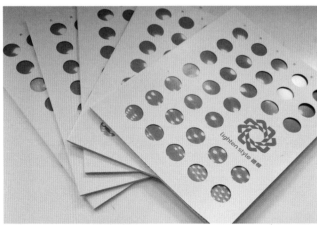

Design Agency:
**Monogum Creative**
Creative Director:
**Alan Lee**
Designer:
**Alan Lee**
Client:
**Hop Shing International Lighting Ltd.**
Photography:
**Alan Lee**
Nationality:
**Hong Kong, China**

设计机构：
Monogum创意设计工作室
创意总监：
阿兰·李
设计师：
阿兰·李
客户：
合成国际灯饰有限公司
摄影：
阿兰·李
国家：
中国，香港

As a newly established company, Lighten Style needed an attractive brochure to sell themselves, especially inside the exhibition booth, there are over thousand of competitors beside their booth. Therefore, an extraordinary cover design is definitely a winning tool for them. The cover design using the circle die-cut holes to create the colour contrast from inside page, it is very eye-catching that can be attracting people from distance. The inside page design uses the minimal approach to avoid overriding the clearness of the important technical information.

作为一个新兴公司，其需要一个十分引人注目的宣传册，从而在拥挤的展亭中脱颖而出，打败其他同类产品。因此，一个匠心独运的封面设计成为打败其他竞争对手的关键武器。封面设计运用圆形模切打孔，使内页中的色调与封面形成鲜明的视觉对比，即使站在远处的人们也会轻易被独特的设计而吸引。内页的设计运用了简约的设计手法，避免干扰重要技术信息的介绍。

# Penedo Typeface

"佩内多字体"宣传册

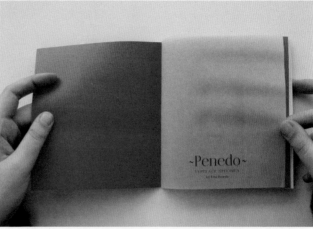

Design Agency:
**Lisa Penedo**
Creative Director:
**Lisa Penedo**
Designer:
**Lisa Penedo**
Photography:
**Bernardo Pinheiro**
Nationality:
**Portugal**

设计机构:
丽莎·佩内多
创意总监:
丽莎·佩内多
设计师:
丽莎·佩内多
摄影:
伯纳德·佩赫罗
国家:
葡萄牙

Penedo is a Typeface made by the designer for a typography school project. The designer's personal taste for contrasting typefaces leads him to this interpretation of a stylish type with a romantic touch and a decorative form. Penedo is the name of an historic place known for its mystical and romantic atmosphere in Coimbra, Portugal, and its also said to be the perfect place for poets. Penedo is also the designer's last name.

佩内多字体是设计师专为一个印刷学校而设计的项目。设计师对对比性字体的独特品位引领其运用浪漫的笔触和装饰性形态将一个独特字形完美诠释。佩内多是葡萄牙科英布拉一个神秘而浪漫的历史胜地,据说该地也是诗人寻找灵感的最佳场所。此外,佩内多也是设计师的姓。

# Typefaces Catalogue 字体型录

Creative Director:
**Lisa Penedo**
Designer:
**Lisa Penedo**
Photography:
**Bernardo Pinheiro**
Nationality:
**Portugal**

创意总监：
丽莎·佩内多
设计师：
丽莎·佩内多
摄影：
伯纳德·佩赫罗
国家：
葡萄牙

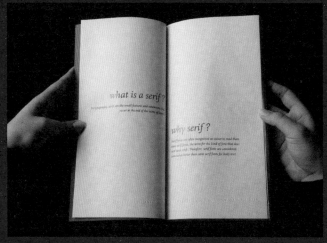

This was a school project for the Typography Class, in which the designer had to select typefaces and organise them by name of typeface, name of type designer, technique, form, historical position and another characteristic of his choice, organising them by type of serif and proper use. The binding was made by the designer, in folders, and he also added instructions to open the perforation to discover the map of the typeface anatomy. The rough yellow paper was extremely important to give life to this catalogue, which was printed with red and black, giving it the purpose for which it was made.

这是专为一个印刷课程而设计的学校项目,设计师需要对字体的名称、字体设计师的名字、设计技巧、形态和历史地位以及字体的额外特性,按照衬线类型和使用属性进行筛选和组织。整个项目采用折页式装订,此外,设计师还在打孔的部分添加了说明,以便使读者更清晰地欣赏到字体剖析示意图。粗糙的黄色纸张巧妙地为这一目录增添了无限活力,红色和黑色色调印刷将该项目的制作意图彰显得淋漓尽致。

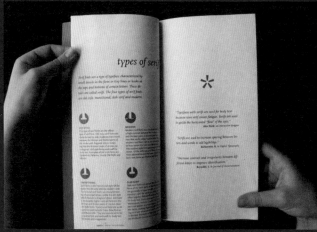

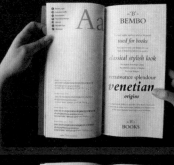

# Sound:Frame Festival Catalogue "声音:构架" 节日型录

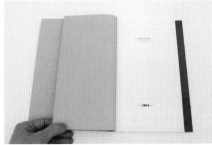

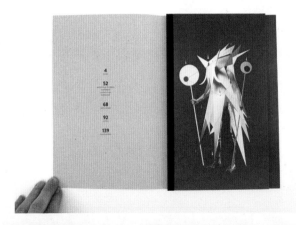

Creative Director:
**Leonhard Lass**
Designer:
**Michael Merzlikar / Martin Embacher / Stephanie Rappl / Vera Svoboda**
Client:
**Sound: Frame**
Photography:
**Andreas Waldschütz & others**
Nationality:
**Austria**

创意总监:
莱昂哈特·赖斯设计工作室
设计师:
迈克尔·墨兹里卡,马丁·埃姆巴切尔,史蒂芬妮·莱普尔,维拉·斯沃博达
客户:
声音:构架节
摄影:
安德烈·瓦尔德库特兹等
国家:
奥地利

Sound:frame is an annual festival for audiovisual expressions in Vienna. The catalogue is the festival's central source of information compiling theory and scientific texts, a presentation of works and performances and a register of all artists involved.

"声音:构架"是维也纳一年一度的视听传达节。该目录中囊括了大量的信息编辑理论和科学文字,是整个节日的广告信息源头,集中展现了所有参与的艺术家的作品和表演。

# Chemisteas

Chemisteas茶叶公司品牌型录

Creative Director:
**Maggie Tsao**
Designer:
**Maggie Tsao**
Client:
**Chemisteas**
Photography:
**Maggie Tsao**
Nationality:
**USA / Taiwan, China**

创意总监：
玛吉·曹
设计师：
玛吉·曹
客户：
Chemisteas茶叶公司
摄影：
玛吉·曹
国家：
美国，中国台湾

It is the brand catalogue for a new fine teas company. The book goes through all aspects of the company including their logo, how to use it, colour palettes, packaging, and corporate stationery.

该项目是专为一个新成立的精品茶叶公司而设计的品牌型录。该书中囊括了公司的所有方面，涉及标识、标识的运用、配色方案、包装以及公司文具的设计等。

# Maya Negri SS11 Press Kit 玛雅·奈格里服饰SS11宣传资料组合

Design Agency:
**Koniak Design**
Creative Director:
**Nurit Koniak**
Designer:
**Nurit Koniak**
Client:
**Maya Negri**
Photography:
**Ran Golani**
Nationality:
**Israel**

设计机构：
肯尼亚克设计工作室
创意总监：
纳里特·肯尼亚克
设计师：
纳里特·肯尼亚克
客户：
玛雅·奈格里服饰
摄影：
兰·格拉尼
国家：
以色列

For the launching of her spring summer 2011 collection, fashion designer Maya Negri comissioned the studio to create a unique press kit, which includes a catalogue, poster and fashion film, directed by Diego Prilusky and Michal Lerman.
The idea behind minimal form - the name of the collection, represented by the gestalt letter M, is to celebrate the clean-cut design of the collection pieces.
The collection, inspired by the designer's journey to Japan, includes solid colours, airy fabrics and loose fits.

　　服装设计师玛雅·奈格里在发布2011年春夏季服装之际，委托肯尼亚克设计工作室为其设计一套独特的宣传资料袋，包括目录、海报以及由迪亚哥·普里卢斯基和米甲·勒曼导演的时装电影设计。
　　设计以简约形态为主，并以该服饰系列的命名为主要目的，运用字母"M"的完整形态以彰显该服饰系列的清晰轮廓。
　　该服饰系列的设计受到了设计师日本之行的启发，纯色色调、轻快的面料和宽大合体的衣着风格是该系列的特色所在。

# Catalogue 型录

Design Agency:
**Zoo Studio**
Creative Director:
**Gerard Calm**
Designer:
**Xavier Castells**
Client:
**Mirplay**
Photography:
**Xavier Castells**
Nationality:
**Spain**

设计机构：
动物园设计工作室
创意总监：
杰拉德·卡尔姆
设计师：
泽维尔·卡斯特
客户：
Mirplay家具公司
摄影：
泽维尔·卡斯特
国家：
西班牙

It is the furniture catalogue for nurseries and primary schools. The designer likes the bright colour and makes it beautiful.

该项目是专为托儿所和小学设计的家具型录。设计师喜欢运用鲜亮的色调以打造美妙的视觉效果。

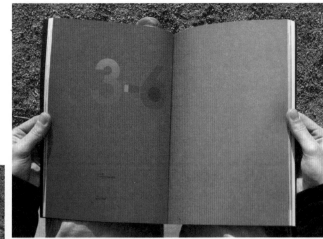

# Branko Kincl - XYZ 布兰科·金科尔-XYZ展览型录

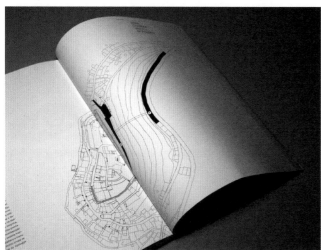

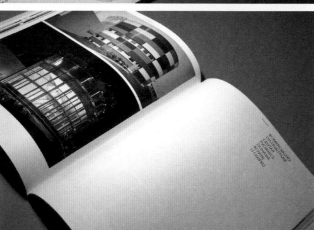

Design Agency:
**Sensus Design Factory Zagreb**
Creative Director:
**Nedjeljko Spoljar**
Designer:
**Nedjeljko Spoljar / Kristina Spoljar**
Client:
**Kabinet grafike HAZU**
Nationality:
**Croatia**

设计机构：
萨格勒布Sensus设计工厂
创意总监：
耐蒂艾尔克·斯颇里扎
设计师：
耐蒂艾尔克·斯颇里扎，克莉斯汀娜·斯颇里扎
客户：
Kabinet grafike HAZU博物馆
国家：
克罗地亚

The name and concept of the exhibition XYZ were designed as a three-dimensional presentation of the author's ideas, thoughts and actions in the field of urban planning and architecture. The authors of three texts give a plastic description of Kincl's work, thus the logical division of the book into the X, Y, Z part, a direct reference to the three-dimensional axis and the iterative principles of working directly in 3D, making the author's approach very specific. Graphic motif of the cross – a typical architectural symbol – is consistent throughout the book, mapping the position of image material, indicating the plan of the layout network that holds a very heterogeneous material together, but also allows for the occasional excesses in the treatment of illustration.

XYZ展览的名称和理念是运用三维形式将作者的创作意图、思想和城市规划以及建筑领域内的行动彰显得淋漓尽致。三篇文字的作品为布兰科·金科尔作品进行了整体的描述，并按照逻辑顺序划分成X, Y, Z三部分，直接用三维形式展现了作品的三维轴向和迭代原理，突出了作者独特的创作手法。十字架的图形主题，作为典型的建筑符号，贯穿整个图书之中，测绘出图案材料的位置，表明网状布局的规划。该规划巧妙地将一个非均质物材料结合在一起，但同时也允许插画设计中的偶尔出血。

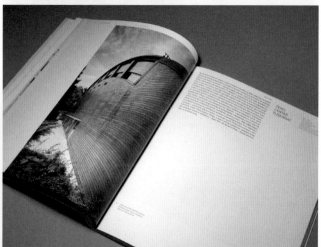

# Spa Catalogue 水疗馆型录

Design Agency:
**Studio Baklazan**
Art Director:
**Urszula Kluz-knopek**
Designer:
**Barbara Łukasik**
Client:
**Comfortum**
Nationality:
**Poland**

It is a project on the design and content of Spa Catalogue. This catalogue is published cyclically by Confortum - a company from Łódź. It presents the best spas in Poland with a short description in 3 languages - with photos. It also contains articles generally concerning the Spa resort topic. It is available with an appendix in which contact information for all the presented spas as well as example prices is included.

该项目是一个有关设计与水疗馆介绍的型录。该型录由罗兹一家名为Confortum的公司定期出版。目录运用三种语言并附以图片，以简约的语言描绘了波兰最佳水疗馆。此外，其中也囊括了有关水疗度假村主题的文章。同时，该目录还在附录中设置了有关书中水疗馆的联系方式和服务价格。

# MeWe

"MeWe" 展览型录

Design Agency:
**Superultraplus Design Studio**
Designer:
**Anders Bergesen / Holger Jungkunz / Arne Schneider**
Client:
**State Academy of Art and Design Stuttgart**
Nationality:
**Germany**

设计机构:
Superultraplus设计工作室
设计师:
安德斯·柏格森, 霍尔格·章戈昆兹, 爱恩·施耐德
客户:
斯图加特国家艺术与设计学院
国家:
德国

"MeWe" is an exhibition catalogue showing student works from the past five years, made in the department of Industrial Design at the State Academy of Art and Design in Stuttgart, Germany. The concept of the catalogue, and the instrument to bind the various projects together, was the visualisation of "the fight within the creative process" through kung-fu images, and quotes by both philosophers and martial artists.

"MeWe"展览型录旨在完美展现德国斯图加特国家艺术与设计学院工业设计专业过去五年的学生作品。该目录的设计理念以及设计手法是将多样化的项目装订在一起，引用功夫形象以及哲学家和武术家的名言形象地诠释"创意过程中的矛盾与冲突"。

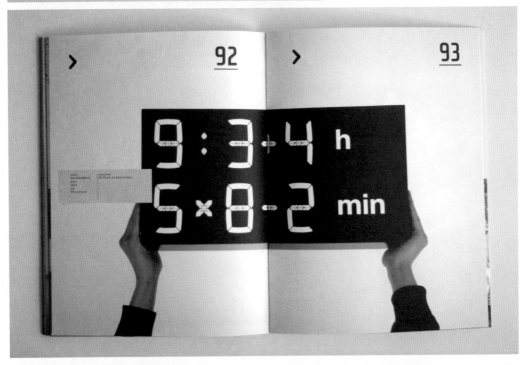

# Schräglage / Incline  "角度/倾向"展览型录

Design Agency:
Superultraplus Design Studio
Designer:
Anders Bergesen / Holger Jungkunz
Client:
Axel Teichmann
Nationality:
Germany

设计机构:
Superultraplus设计工作室
设计师:
安德斯·柏格森，霍尔格·章戈昆兹
客户:
阿克塞尔·特其曼
国家:
德国

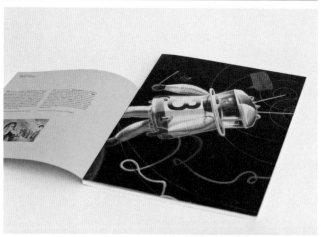

"Schräglage / Incline" is an exhibition catalogue for the German artist Axel Teichmann. Modern humanity forms the focus of Axel Teichmann's paintings. Surrounded by technical devices, graphic or virtual elements and curious objects, he struggles for control of the world he has created.
The catalogue is bound with a soft cover and dust-jacket. The introduction is printed on blue paper, showing the preparation and hang-up of the paintings. The second part shows the paintings on display.

"角度/倾向"展览目录专为德国设计师阿克塞尔·特其曼而设计。现代人性化设计风格巧妙地为阿克塞尔·特其曼的绘画作品制造了焦点所在。技术装置、图形或虚拟元素以及奇异物件包裹下的作品展现了设计师为其打造的世界所进行的不懈努力。

该目录设有平装封面和防尘书套。简介部分以蓝色纸张为原料，展现了绘画作品的准备工作和设计中遇到的困难。第二部分所展示的是展览画作。

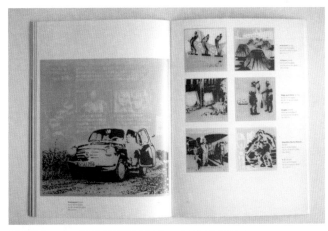

# Toegepast 14    第十四届Toegepast博览会型录

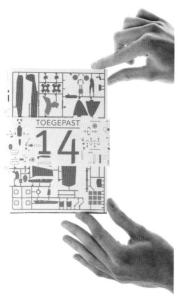
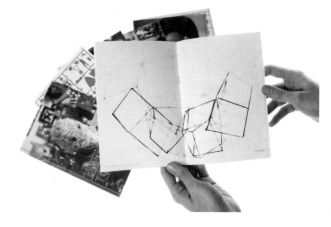

Designer:　　　　　设计师:
**Jens Dawn**　　　　延斯·潼恩
Client:　　　　　　　客户:
**Design Platform Limburg**　林堡设计论坛
Photography:　　　摄影:
**Liesje Reyskens**　林赛·赖斯肯斯
Nationality:　　　　国家:
**Belgium**　　　　　比利时

It is the catalogue of the 14th exposition in Z33 "artcentre" Hasselt by young designers that won the Design Platform Limburg contest. This catalogue is divided in 5 books (one for every designer) and 1 book with information about the contest and Design Platform Limburg. The theme is inspired by the buildingkit of airplanes, but in a more abstract form.

该项目是专为曾经获得林堡设计论坛大赛的设计师在哈瑟尔特市"艺术中心" Z33画廊所举办的第十四届博览会而设计的型录。该型录包括5本设计师图书（每本介绍一个设计师）以及1本以介绍大赛和林堡设计论坛相关信息的图书。设计的主题受到了飞机拼图的启发，以一种更为抽象的形式将这一理念诠释得淋漓尽致。

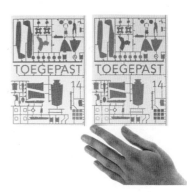

# Untitled Untitled Untitled  "无题"型录

Creative Director:
**Thorleifur Gunnar Gíslason**
Designer:
**Thorleifur Gunnar Gíslason**
Client:
**Living Art Museum in Iceland**
Photography:
**Thorleifur Gunnar Gíslason**
Nationality:
**Iceland**

创意总监：
拖雷夫·贡纳尔·吉斯拉森
设计师：
拖雷夫·贡纳尔·吉斯拉森
客户：
冰岛生活艺术博物馆
摄影：
拖雷夫·贡纳尔·吉斯拉森
国家：
冰岛

Untitled Untitled Untitled is an exhibition catalogue and a poster for a fictional exhibition at the Living Art Museum (Nylo) in Reykjavík, Iceland. The exhibition would cover the graphics works of Dieter Roth.

The result is a clean cut catalogue / folder for general usage for the Living Art Museum and a poster for the exhibition that works as a wrapping paper as well for the catalogue.

该项目是专为冰岛雷克雅未克生活艺术博物馆中举办的虚拟展览而设计的展览型录和海报。该展览中囊括了设计师迪特尔·罗斯的平面创作作品。

最终的设计方案是为冰岛生活艺术博物馆打造了一个简洁、轮廓鲜明的目录或折叠式印刷品以及一个专为展览而设计并可以充当目录包装纸的海报。

# Product Catalogue 产品型录

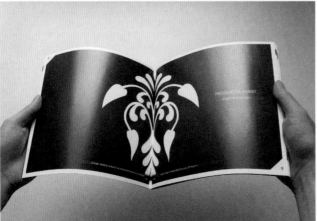

Design Agency:
**SZMER**
Creative Director:
**Jakub Ben**
Designer:
**Jakub Ben**
Client:
**Avant [Advertising ceramics]**
Photography:
**Jakub Ben**
Nationality:
**Poland**

设计机构:
SZMER设计工作室
创意总监:
雅各布·本
设计师:
雅各布·本
客户:
阿万特[陶瓷制品广告宣传]
摄影:
Jakub本
国家:
波兰

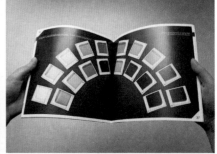

After visited factory of the porcelain, the designer decided to make the catalogue shine with fresh, glossy colours and combine it with classic patterns. He has putted products photos into frames similar to polaroid, which was quite risky. But the effect was very interesting, and even today is his reminder that it's worth to experiment with your projects, even when you are trying to combine seemingly unsuitable elements.

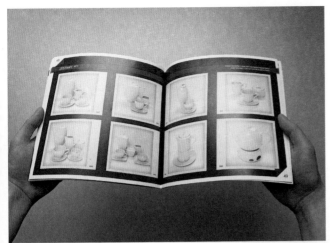

设计师在参观陶瓷制品加工工厂之后，决定打造一个清新独特、颜色鲜艳，并以经典图案为主题的产品型录。设计师巧妙地将产品照片运用到类似人造偏光板的画面中，匠心独运，可谓是一个突破性设计。最终的设计结果表明，这一举措是成功的，即使到了现在，设计师依然认为其他项目也可以尝试运用该手法，甚至是添加一些看起来似乎不太合适的元素。

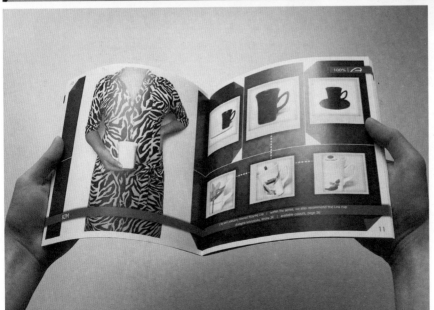

# Nike - Train to Win "耐克—练战英雄"型录

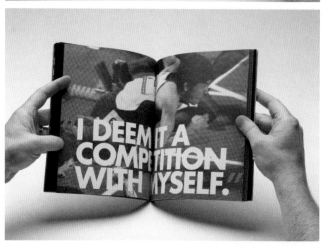

Design Agency:
**PLAZM**
Art Director:
**Joshua Berger**
Designer:
**Joshua Berger / Ian Lynam**
Writer:
**Tiffany Lee Brown**
Client:
**Nike Asia Pacific**
Nationality:
**USA**

设计机构:
PLAZM设计工作室
艺术总监:
约书亚・柏格
设计师:
约书亚・柏格,伊恩・莱纳姆
文档编写:
蒂凡尼・李・布朗
客户:
耐克公司亚太区
国家:
美国

A 90+ pages media kit features the training regimens and stories of six key Nike athletes in the run-up to the Beijing Olympics.

该项目是一个90多页的媒体资料包,以介绍六位主要耐克运动员为备战北京奥运会所进行的训练以及训练中所发生的故事为主题。

# Nike Sisters - US Media Kit

"耐克女子系列运动鞋 – 美国媒体资料包"型录

Design Agency:
**PLAZM**
Art Director:
**Joshua Berger**
Designer:
**Thomas Bradley**
Client:
**Nike**
Photography:
**Allen Benedikt**
Nationality:
**USA**

设计机构：
PLAZM设计工作室
艺术总监：
约书亚·柏格
设计师：
托马斯·布拉德利
客户：
耐克公司
摄影：
艾伦·本耐迪克特
国家：
美国

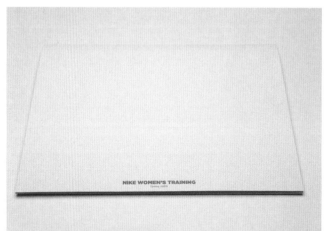

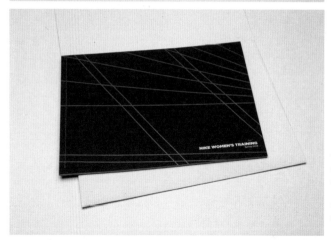

PLAZM created an oversized book for the launch of the Nike's Sisters training line in the US. The story emphasizes the latest technological innovations associated with the Sisters line of footwear. Plazm also designed letter-pressed/foil stamped invitations, and a complete English/Spanish media site tons of downloadable product images.

PLAZM设计工作室为美国耐克女子训练产品系列而设计的大型图书。该故事体现了女子运动鞋全新的技术革新。此外，PLAZM设计工作室还设计了字母压花/烫金邀请函以及一个囊括大量可下载产品图片的完整英语/西班牙语媒体网站。

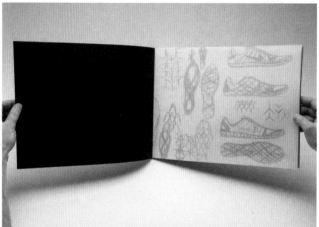

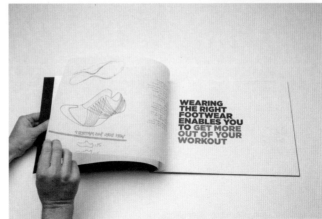

# MCAD CGD Catalogue  MCAD CGD型录

Designer:
**Matt Van Ekeren / Molly Hein / Catherine Grothe Bicknell / Christina Rimstad**
Photography:
**Patrick Kelley / Erin Nicole Johnson**
Nationality:
**USA**

设计师:
马特·凡·埃克林，莫莉·海因，凯瑟琳·葛罗斯·毕克奈尔，
克里斯蒂娜·利姆斯戴德
摄影:
帕特里克·凯利，艾玲·妮可·约翰逊
国家:
美国

This design takes the traditional catalogue to another level, engaging the viewer with an elaborate folding system that nestles the admission booklet within informational take-away cards.

该设计巧妙地将传统型录上升到一个更高的层次，引领读者共同欣赏一个匠心独运的折叠系统，信息化可携带卡片将入门手册囊括其中。

# Toegepast 15 "De meeting" 第十五届Toegepast "博览会" 型录

Designer:
**Jens Dawn**
Client:
**Design Platform Limburg**
Photography:
**Filip van Roe**
Nationality:
**Belgium**

完成时间：
延斯·潼恩
客户：
林堡设计论坛
摄影：
菲力普·凡·罗伊
国家：
比利时

It is the catalogue of the 15th exposition in Z33 "artcentre" Hasselt by young designers that won the Design Platform Limburg contest. The catalogue is also a magazine and the overall theme of the expo is mirrors, titles in mirror and transparent colours.

该项目是专为曾经获得林堡设计论坛大赛的设计师在哈瑟尔特市"艺术中心" Z33画廊所举办的第十五届博览会而设计的型录。该目录本身也是一本杂志，该展览会的全部主题是镜面反射、镜面标题以及透明色彩的设计。

# Atelier Sedap 2011

2011 Atelier Sedap公司产品名录

Design Agency:
**Ludovic Roth Design Studio**
Creative Director:
**Fabrice Berrux / Benoît Bohu**
Designer:
**Ludovic Roth**
Client:
**Atelier Sedap**
Photography:
**Ludovic Roth / Fabrice Berrux**
Nationality:
**France**

设计机构：
卢多维克·罗斯设计工作室
创意总监：
法布里斯·伯卢克斯，贝诺特·波胡
设计师：
卢多维克·罗斯
客户：
Sedap工作室
摄影：
卢多维克·罗斯，法布里斯·伯卢克斯
国家：
法国

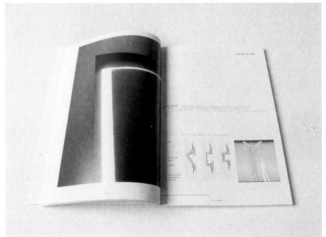

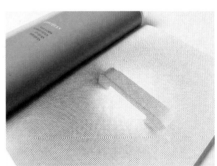

Atelier Sedap is a French lighting manufactured based in Nantes. Creativeness and R&D at Atelier Sedap open new possibilities for the use of plaster, and especially new solutions for recessed lighting. For the 2011 annual catalogue, the designers realised a rebranding to boost visual communication and then created the new visual identity: logo, annual catalogue, business cards, invoices, etc...The goal was to have an overall presentation clear, simple and elegant.

法国Atelier Sedap公司以生产照明设备为特色，总部坐落于法国南斯市。Sedap工作室的卢多维克·罗斯创意团队开创了石膏全新运用的先河，并为壁灯的设计开发了崭新的设计理念。对于该公司的2011年度产品型录，设计师对品牌进行重塑以后大大提升了视觉传达的力量，并创建了展现的视觉识别系统，涉及标识、年度目录、名片、发票等的设计。设计的目标是清晰、简约、优雅地传达出产品的个性与魅力。

# Dix Heures Dix  Dix Heures Dix 公司型录

Design Agency:
**Ludovic Roth Design Studio**
Creative Director:
**Fabrice Berrux / Benoît Bohu**
Designer:
**Ludovic Roth**
Client:
**Dix Heures Dix**
Photography:
**Ludovic Roth / Fabrice Berrux**
Nationality:
**France**

设计机构：
卢多维克·罗斯设计工作室
创意总监：
法布里斯·伯卢克斯，贝诺特·波胡
设计师：
卢多维克·罗斯
客户：
Dix Heures Dix公司
摄影：
卢多维克·罗斯，法布里斯·伯卢克斯
国家：
法国

Dix Heures Dix's show-room and production facility are based in the West of France, the Nantes area – close to the Atlantic Ocean. Since 2002, Dix Heures Dix has been developing design lighting fixtures. The concept of the Annual Catalogue 2010 is to place each product in different graphic universe that realised on computer.

　　Dix Heures Dix 公司坐落于法国西部南斯地区，靠近大西洋。该项目旨在展现该公司的经典产品和生产设备。自2002年以来，该公司一直致力于照明器材的开发设计。2010年度型录的设计理念是将每个产品放置到不同的平面领域之中，并运用电脑进行多样化设计。

# Best books from Lithuania / 31 Migliori libri Lituani / 30 Beste Bücher aus Litauen

立陶宛畅销书书目

**Design Agency:**
PRIM PRIM
**Creative Team:**
Migle Vasiliauskaite,
Kotryna Zilinskiene
**Nationality:**
Lithuania

设计机构:
PRIM PRIM设计公司
创意团队:
米格尔·瓦斯里奥斯凯特、
科特丽娜·泽林斯基尼
国家:
立陶宛

The three catalogs represent best Lithuanian writers in English, Italian and German. Each catalog has its main active colour: German – ink blue, Italian – basil green, English – neon pumpkin.

三本书目分别以英文、意大利文和德文呈现了立陶宛的畅销书。每本书目都有其独特的色彩：德文版采用了墨水蓝，意大利文版采用了罗勒绿、英文版采用了霓虹橙。

# Sheridan Illustration Graduate Show
谢里登插画设计专业毕业生作品展型录

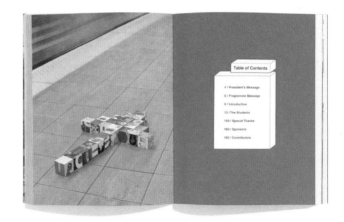

Design Agency:
Rethink
Creative Director:
Ian Grais / Chris Staples / Jeff Harrison
Designer:
Kim Ridgewell
Client:
Sheridan Institute of Technology and Advanced Learning
Photography:
Kim Ridgewell / Rory O'Sullivan / Leia Rogers / John Jones
Nationality:
Canada

设计机构:
新理念设计工作室
创意总监:
伊恩·格雷斯,克里斯·斯台普斯,杰夫·哈里森
设计师:
吉姆·里奇韦尔
客户:
谢里登技术学院
摄影:
吉姆·里奇韦尔,罗理·奥沙利文,莱亚·罗杰斯,约翰·琼斯
国家:
加拿大

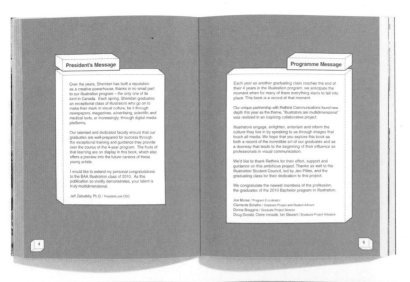

The theme of Sheridan's 2010 Illustration Grad Show was "Illustrators are multidimensional." For the students' final project, they each created six specific illustrations and applied them to the sides of a 3-D cube. The cubes were then grouped together to form larger shapes, images and statements about the potential of illustration.

谢里登插画设计专业2010年毕业生作品展的主题是"多维的插画师"。在毕业设计的过程中,每个毕业生创作六个别具一格的插画作品,并最后将它们放置到一个三维立方体的各个侧面之中。最终,这些立方体经组合之后所形成的大型几何形状、图案和状态完美展现了插画设计的潜力。

# Catalogue 50 Years - Aleko Konstantinov Satirical Theatre

阿列科·康斯坦丁诺夫讽刺剧院50周年型录

**Design Agency:**
Noblegraphics Creative Studio
**Creative Director:**
Chavdar Kenarov
**Designer:**
Miroslav Krustev
**Client:**
Aleko Konstantinov Satirical Theatre
**Photography:**
Atanas Kanchev
**Nationality:**
Bulgaria

设计机构：
Noblegraphics创意设计工作室
创意总监：
恰夫达尔·柯那洛夫
设计师：
米洛斯拉夫·克鲁斯特夫
客户：
阿列科·康斯坦丁诺夫讽刺剧院
摄影：
阿塔纳斯·肯彻夫
国家：
保加利亚

The catalogue is for the 50th anniversary of Aleko Konstantinov Satirical Theatre founding. It recounts the history of the theatre from 1957 until today. The selected photographs represent the best plays that have been performed there.

该项目是专为阿列科·康斯坦丁诺夫讽刺剧院成立50周年而设计的型录。该目录再现了剧院从1957年创办至今的发展轨迹。精选的照片展现了在该剧院中上演的精彩演出。

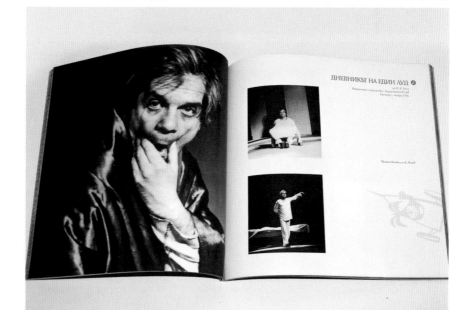

# Lexon Design Book "莱克森公司设计图书"型录

Design Agency:
**Ludovic Roth Design Studio**
Creative Director:
**Fabrice Berrux / René Adda**
Designer:
**Ludovic Roth**
Client:
**Lexon**
Photography:
**Olivier Mesnage**
Nationality:
**France**

设计机构：
卢多维克·罗斯设计工作室
创意总监：
法布里斯·伯鲁克斯、莱内·阿达
设计师：
卢多维克·罗斯
客户：
莱克森公司
摄影：
奥利维尔·麦斯内奇
国家：
法国

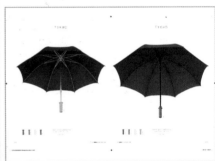

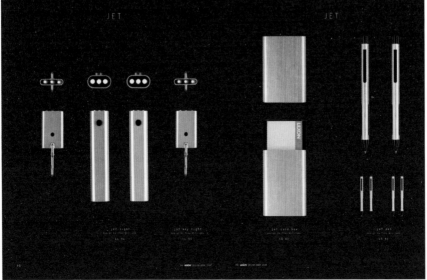

Lexon is a French objects manufactured based in Boulogne- Billancourt in France. The main idea was to have a minimal presentation, simple and elegant, to emphasize all the different collection of watches, clocks and office stuff.

坐落于法国布洛尼–比扬古市的莱克森公司以加工法国特色产品为特色。该设计的主要目标是打造一个精致、简约、高雅的图书形象，突出多样化的手表、时钟和办公用品种类。

# Parad

Parad服饰精品店型录

Design Agency:
**Communication Bureau Proekt**
Creative Director:
**Roman Krikheli**
Designer:
**Dmitry Rybalkin / Andrey Ilyaskin / Dmitry Fedorov**
Client:
**Parad**
Nationality:
**Russia**

设计机构：
Proekt传达设计工作室
创意总监：
罗马·克里克赫立
设计师：
德米特里·莱贝尔金、
安德烈·伊来斯金、
德米特里·费德洛夫
客户：
Parad服饰精品店
国家：
俄罗斯

Parad is a Russian fashion boutique with a friendly face and a bit glamorous image. First, the designers performed a rebranding to boost visual communication even more, then they created a number of catalogues: work books and books for advertising. This is the first branding catalogue for Parad made in focus of the new style. The concept of the new catalogue is called "Alice in Wonderland". All lots are shot in stylised mirror boxes, a few spreads bear creative illustrations and counterintuitive paragraphs inspired by Lewis Carroll.

坐落于俄罗斯的Parad服饰精品店拥有一个亲切的外观和迷人的形象。首先，设计师进行了一系列的品牌重塑活动，大大推动了视觉传达的力量；随后，设计师创建了若干目录，涉及工作手册和广告宣传图书。该项目是专为Parad服饰精品店设计的首个品牌目录，全新的设计风格是整个设计的焦点所在。该全新目录的设计理念围绕"艾利斯梦游仙境"展开。所有的商品均以风格独具的反光镜箱为拍摄背景，部分跨页设置了创意插画和反直觉照片，该设计灵感受到了刘易斯·卡罗尔的启发。

# CM2  CM2建筑设计处型录

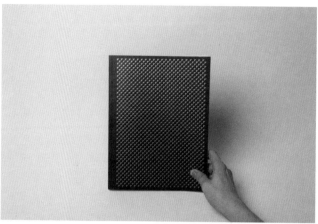

Design Agency:
**Communication Bureau Proekt**
Creative Director:
**Roman Krikheli**
Designer:
**Andrey Koodenko / Alexander Blükher / Yury Novenkov / Ksenia Polyakova**
Client:
**SM2 Lestnitsy**
Nationality:
**Russia**

设计机构：
Proekt传达设计工作室
创意总监：
罗马·克里克赫立
设计师：
安德烈·库登科，亚历山大·布鲁克，郁理·诺万柯夫，肯塞尼亚·伯利亚科瓦
客户：
CM2建筑设计处
国家：
俄罗斯

CM2 is an architectural bureau that specialises in constructing stairs for premium private and commercial buildings. The designers have chosen the most spectacular examples of their works and created unusually-looking images with collage technique. Then they added some aesthetic typography and designed two versions of cover: a regular printed edition and an exclusive wooden laser-cut casing.

　　CM2建筑设计处一直致力于为高级私人住宅和商业办公建筑提供楼梯的设计服务。设计师精选了该设计处的经典项目，运用抽象拼贴画的技法打造了非凡的品牌形象。随后，设计师添加了一些美学印刷字体，并设计了两个版本的封面：一个是常规的印刷版，而另一个则是独特的木质激光切割包装外壳。

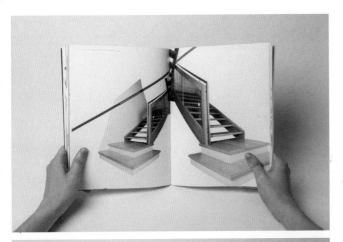

# Nicole LaFave Lookbook 妮可·拉法夫型录

Creative Director:
**Nicole LaFave**
Designer:
**Nicole LaFave**
Nationality:
**USA**

创意总监：
妮可·拉法夫
设计师：
妮可·拉法夫
国家：
美国

This was designed as a promotional hand out extending beyond the typical formats you often find a brochure in. The brochure was set up as individual cards that come contained in a sleeve with a belly band. Each card had different information about Nicole LaFave's Footwear Design experience and showcased some of her portfolio and footwear sketches. The brochure was used to make first contacts with professionals in Italy at Linea Pelle and Italian footwear factories.

该项目是一个设计独特的促销宣传品，与人们经常在宣传手册中看到的截然不同。该手册巧妙地将独立的宣传卡片运用一个腰封完美地衔接在一起，并在外面加以书套。每个卡片上介绍了关于妮可·拉法夫鞋类设计体验的不同信息，并展现了其设计作品和鞋类素描。该手册的制作在妮可·拉法夫与意大利琳琅沛丽皮革展中的专业人士和鞋类加工厂的首次合作中扮演了重要角色。

# Parallels. Lithuanian Architecture: Three Eras, Three Faces

"平行——立陶宛建筑：三个时期，三个方面"展览型录

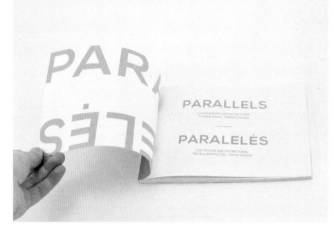

**Design Agency:**
PRIM PRIM
**Creative Team:**
Migle Vasiliauskaite,
Kotryna Zilinskiene
**Nationality:**
Lithuania

设计机构：
PRIM PRIM设计公司
创意团队：
米格尔·瓦斯里奥斯凯特、
科特丽娜·泽林斯基尼
国家：
立陶宛

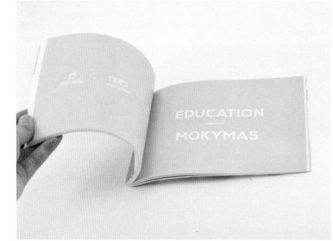

This project includes catalog and posters design for the exhibition of Lithuanian Architecture. The exhibition took place at the Royal Institute of British Architects (RIBA), London. PRIM PRIM designed three icons representing three different architectural eras. Categories of buildings are distinguished by pastel colour tones.

该项目是为立陶宛建筑展所设计的型录。展览由英国皇家建筑师协会在伦敦举办。PRIM PRIM设计公司设计了三个代表不同时期的图标。在型录中，不同的建筑类型通过不同的色彩区分开来。

# Corporate Identity "Simonswerk"

"Simonswerk" 企业识别型录

Design Agency:
804© Agentur fuer visuelle Kommunikation
Creative Director:
Oliver Henn / Helge Rieder
Designer:
Oliver Henn / Helge Rieder / Carsten Prenger
Client:
Simonswerk GmbH
Nationality:
Germany

设计机构:
804c视觉传达设计工作室
创意总监:
奥利弗·海恩, 黑尔格·里德
设计师:
奥利弗·海恩, 黑尔格·里德, 卡斯腾·普莱戈
客户:
Simonswerk股份有限公司
国籍:
德国

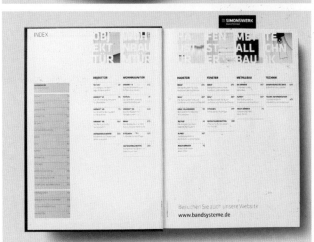

In the course of redesign, all communication media, from letterhead, over website to the general catalogue, were optimised. After the 8-month realisation phase of general catalogue, the official corporate design relaunch took place on the trade fair fensterbau/frontale in Nuremberg. The catalogue, approx. 550 pages thick, is a key medium of the product and enterprise communication and was also fully reworked by 804©. The design office not only enhanced the optical surface in terms of visual consistency, but also simplified the contents, the contents structure, thereby significantly increasing the utility value.

在再设计的过程中,所有的传达媒介,包括信头、网站和目录等全部得到了重新优化。

在对目录进行8个月的设计之后,正式的企业设计在德国纽伦堡窗口/前言交易会上重新发布。该目录将近550页,是产品和企业传达的一个关键部分,全部由804设计工作室重新加工。该工作室不仅在视觉的一致性方面强化了光学表面,同时也对内容进行了简化,从而有效地增强了目录的使用价值。

# Silly Things

"愚蠢的事"展览型录

It is 80-page perfectly bound full colour catalogue for Silly Things, an exhibition at Fold Gallery in London. The exhibition explored the depiction of everyday ephemera or the body, and the humour often inherent in the process. Text was illustratively set at angles and skewed to give the impression of depth and support the focus on tangible objects, and titles are split vertically over spreads to make the design more playful.

这个装订精美的80页全色目录专为在伦敦折叠画廊举办的名为"愚蠢的事"展览而设计。这一展览探讨了人们日常所见或自身发生的荒唐事件以及期间所存在的幽默气息。文字呈一定角度排列,倾斜的形态给人留下一种深邃的印象,并将读者的注意力聚焦在实体之上,跨页中的标题采用垂直分布的形式,巧妙地为设计增添了无限情趣。

Design Agency: **Bellamy Studio**
Creative Director: **Andrew Bellamy**
Designer: **Andrew Bellamy**
Client: **Fold Gallery London**
Photography: **In house**
Nationality: **USA**

设计机构: 贝拉米设计工作室
创意总监: 安德鲁·贝拉米
设计师: 安德鲁·贝拉米
客户: 伦敦折叠画廊
摄影: In house摄影工作室
国家: 美国

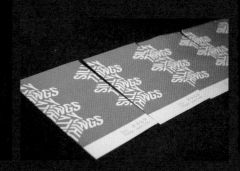

# Honor Oak Church "贵橡教堂"宣传册

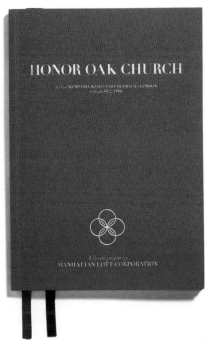

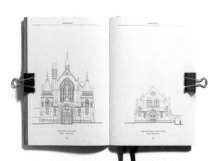

Design Agency:
**Exposure**
Creative Director:
**Clare Styles**
Designer:
**Steven Camp**
Client:
**Manhattan Loft Corporation**
Photography:
**Valerie Bennet**
Nationality:
**UK**

设计机构：
Exposure设计工作室
创意总监：
克莱尔·斯戴尔斯
设计师：
史蒂文·坎普
客户：
曼哈顿阁楼公司
摄影：
瓦莱丽·本耐特
国家：
英国

Manhattan Loft Corporation is one of the UK's most innovative property companies, pioneering the idea of loft-living in London. Honour Oak Church is a residential development and refurbishment of a baptist church.
The brochure referenced the bible through subtle design features and production techniques whilst the photography incorporated heroic, religious beams of light.

曼哈顿阁楼公司是英国最富有创新精神的物业公司之一，是伦敦阁楼居住理念的开拓者。贵橡教堂是一个住宅开发项目，其前身是浸信会教堂。
该手册巧妙地运用精致的设计特色和制作技巧，与《圣经》建立起微妙的联系，同时在摄影图片中添加神圣、庄重的光束以彰显自然、端庄之感。

# Burgos 2016 "2016年布尔格斯"宣传册

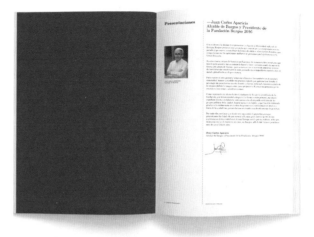

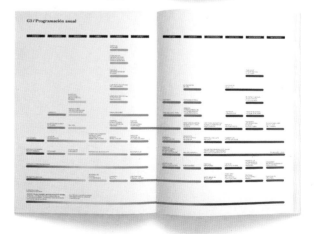

Design Agency:
Erretres
Designer:
Erretres
Client:
City of Burgos, Spain
Nationality:
Spain

设计机构：
厄莱特拉斯设计工作室
设计师：
厄莱特拉斯
客户：
西班牙，布尔格斯市
国家：
西班牙

The city of Burgos presented their candidature as the European Capital of Culture for 2016. Erretres designed their presentation dossier and is currently working on creating their brand, including its logo, design pieces to be used as promotional material as well as environmental graphics, merchandise, among others.

该项目专为西班牙布尔格斯市竞选2016年欧洲文化首都而设计。厄莱特拉斯设计工作室提供简介资料的设计，目前也正在进行品牌的设计工作，包括标识、作为宣传材料使用的设计模块以及环境平面和其他商品设计。

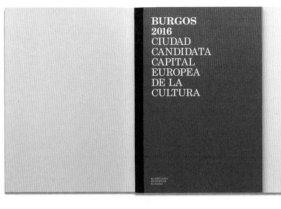

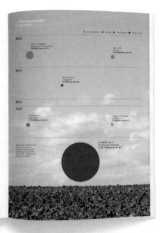

# Cupcake Identity

杯形蛋糕水疗中心&育婴所识别设计手册

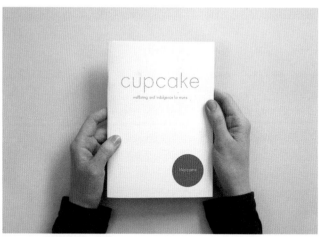

Design Agency:
**Mind Design**
Creative Director:
**Holger Jacobs**
Art Director:
**Craig Sinnamon**
Client:
**Cupcake**
Nationality:
**UK**

设计机构：
思想设计工作室
创意总监：
霍尔格·雅各布斯
艺术总监：
克雷格·赛纳蒙
客户：
杯形蛋糕水疗中心&育婴所
国家：
英国

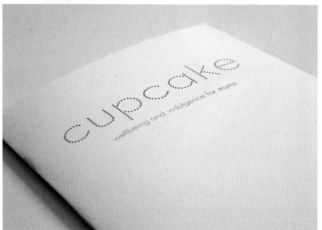

It is the design of identity for Cupcake, a spa and crèche for mothers with young babies. The idea is that mothers can leave their toddlers in the crèche where they are looked after while they enjoy their time in the spa, exercise in the gym or meet for a chat in the café. The challenge was to develop a logo that relates equally to mothers and children. The flexible logo made up from dots can either be used in a single colour or in multiple colours to appeal to children. The multi-coloured dots appear in the interior design as well as in all printed material. Occasionally individual dots were replaced by small images, for example on the wall paper in the play area or on the inside cover of the main brochure.

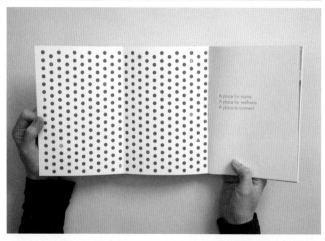

杯形蛋糕水疗中心&育婴所专为妈妈和幼儿们而设计，该项目是专门为其设计的识别体系。在这里，妈妈们能够在体验水疗服务、健身或在咖啡馆中会友的同时，放心地将孩子们托管给该公司，并享受贴心的照顾，而这也恰恰是该设计的核心理念所在。设计的挑战是如何开发一个同时满足妈妈和孩子心理要求的标识。由圆点构成的标识，设计灵活，可适用于淡色或多色印刷，以吸引孩子们的目光。其中，多色圆点主要运用在室内设计以及所有的印刷材料之中。个别的圆点被小图案所取代，例如游戏区的墙纸上以及主要宣传手册的内封之中。

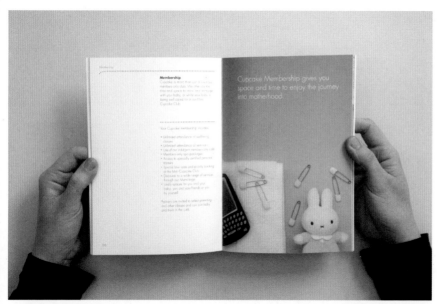

# Getxophoto   Getxophoto国际摄影节宣传册

Design Agency:
**Barfutura**
Designer:
**Sergio Gonzalez Kuhn**
Client:
**Getxophoto**
Nationality:
**Spain**

设计机构:
Barfutura设计工作室
设计师:
塞尔吉奥·冈萨雷斯库恩
客户:
Getxophoto国际摄影节
国家:
西班牙

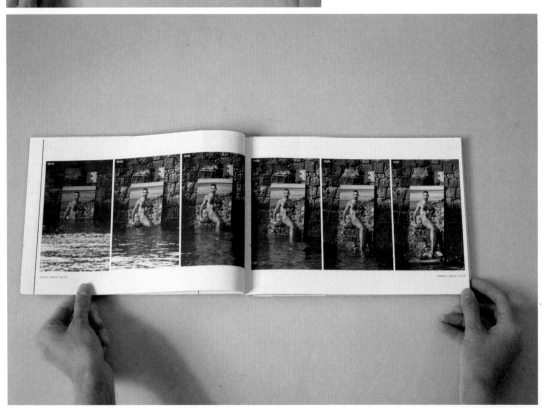

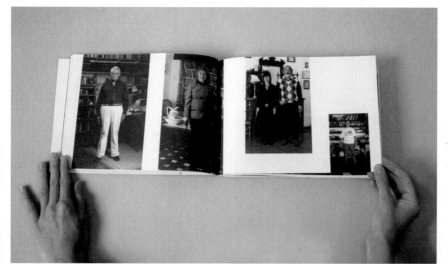

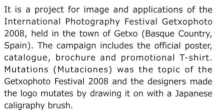

It is a project for image and applications of the International Photography Festival Getxophoto 2008, held in the town of Getxo (Basque Country, Spain). The campaign includes the official poster, catalogue, brochure and promotional T-shirt. Mutations (Mutaciones) was the topic of the Getxophoto Festival 2008 and the designers made the logo mutates by drawing it on with a Japanese caligraphy brush.

该项目是专为2008年在西班牙巴斯克郡Getxo市举办的Getxophoto国际摄影节而设计的形象和应用程序。该活动涉及官方海报、目录、小册子和促销T恤衫的设计。"转变"是2008年摄影节的主题，因此，设计师巧妙地运用日式毛笔对标识进行了独到的"改变"。

# Inside-Exposure "内在" – Exposure综合广告公司宣传册

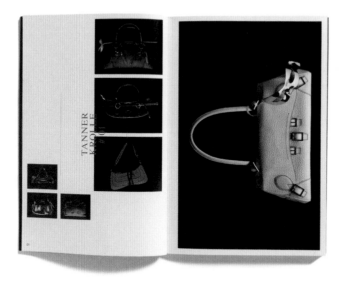

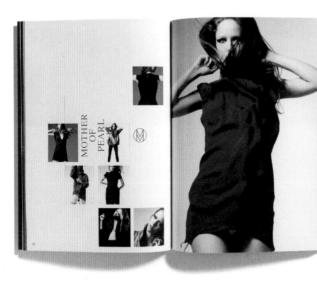

Exposure is a leading communications agency, producing iconic publicity for some of fashions most interesting brands. This brochure tells the story of the design studio within the agency and its work for many of the same clients. Inside each perforated French-fold lies an iconic piece of work produced by the studio.

Potential clients are able to discover the contents within, including a number of additional tipped in pages.

Exposure综合广告公司是一支优秀的广告制作队伍,旨在为时尚品牌提供形象宣传设计。该手册介绍了该公司旗下的设计工作室以及他们与一些客户的合作项目。每一个穿孔法式折页中均设置了一个代表该工作室创作作品的形象。

潜在客户能够清晰地从这一手册中找到信息,其中,页面中还额外设置了若干温馨贴士。

Design Agency:
**Exposure**
Creative Director:
**Simon Shaw**
Designer:
**Steven Camp**
Nationality:
**UK**

设计机构:
Exposure设计工作室
创意总监:
西蒙·肖
设计师:
史蒂文·坎普
国家:
英国

# Coordinated Image

"和谐的形象"宣传册

Design Agency:
**Zoo Studio**
Client:
**Famgoca**
Nationality:
**Spain**

设计机构：
动物园设计工作室
客户：
Famgoca公司
国家：
西班牙

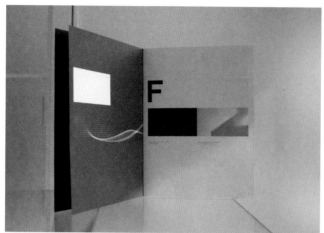

The design includes corporate image and brand for an industrial engineering company. The brand is designed using elements of the enginery field, such as the millimetric grid. The typography also reminds the one used in the software that engineers use. The composition of the typography and the relation with the graph grid gives great personality and modernity to the brand.

该项目是专为一个工业工程公司而设计的企业形象和品牌塑造方案。该品牌的设计大量运用了机械领域的元素，例如毫米网格等。醒目的字体令人自然联想起工程师在软件中常用的字体。字体的组合以及其与图表网格的联系为该品牌创造出独有的魅力和全新格调。

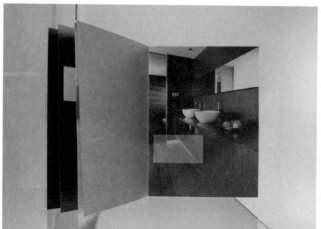

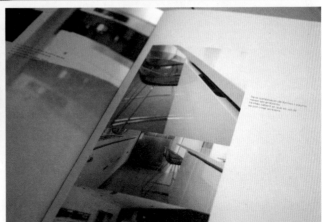

# Internship Log "实习日志" 手册

Design Agency:
**Elke Broothaers**
Client:
**Sint-Lukas Hogeschool Brussel**
Nationality:
**Belgium**

设计机构:
艾尔克·布鲁蒂厄斯
客户:
布鲁塞尔圣·卢卡斯学院
国家:
比利时

This is a logbook of the internship of Elke Broothaers at MilkandCookies. MilkandCookies is a succesfull design agency located in the heart of Brussel. The logbook is filled with all the projects in which she participated during her internship. The frontcover shows a portrait of her and the backcover has the same shot but from behind. It's a symbolic picture that respresents the end of her internship at the end the book.

该项目是设计师艾尔克·布鲁蒂厄斯在MilkandCookies设计公司实习期间所记录的实习日志。MilkandCookies设计公司坐落在比利时首都布鲁塞尔市中心,因设计诸多成功作品而闻名。该日志中展示了设计师在实习期间所参与的所有项目。封面设置了设计师的自画像,而封底则是设计师的背影肖像。该形象巧妙地暗示了该日志的完结篇即象征着设计师实习期的结束。

# 100th Anniversary of Audi in Hungary

"匈牙利奥迪公司成立100周年"宣传册

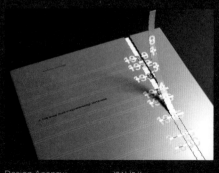

Design Agency:
Studio Borzák
Creative Director:
Márton Borzák
Designer:
Márton Borzák, Réka Diósi
Client:
k3 corp.
Nationality:
Hungary

设计机构：
布罗扎克设计工作室
创意总监：
马顿·布罗扎克
设计师：
马顿·布罗扎克，雷卡·迪奥赛
客户：
K3公司
国家：
匈牙利

It is the brochure designed for the 100th Anniversary of Audi in Hungary. By opening the brochure the viewer gets a brief overview of the most important years in the company's history.

该项目是专为匈牙利奥迪公司成立100周年而设计的纪念画册。打开该画册，读者将能够对该公司自创办以来所经历的大事记进行简单、清晰的了解。

# Heineken Brochure 喜力品牌宣传册

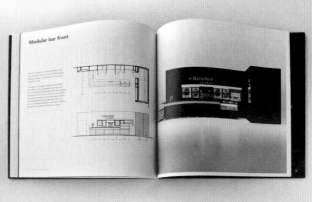
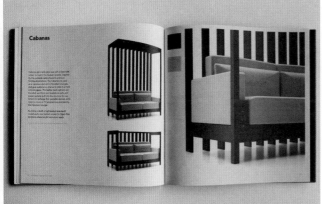

**Design Agency:**
UXUS
**Client:**
Heineken
**Nationality:**
USA

设计机构：
UXUS设计工作室
客户：
喜力公司
国家：
美国

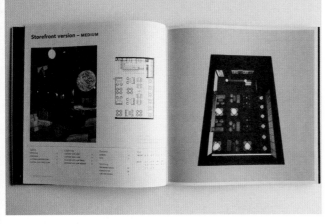

The Heineken Lounge brand brochure captures the essence of the food & beverage concept's design intent, purpose and the application of Heineken's brand equity at airport locations worldwide. Intended as a manual for guiding the roll-out of the modular bar and lounge, the brochure outlines every component and is illustrated with fully rendered 3D layouts for concessionaires and operators to implement a "brand-right" location.
The brochure's design and layout also follow the concept's look and feel. The cover features UV gloss printing on a matte black stock, introducing the Lounge interior's custom black-on-black wallpaper.

　　喜力酒吧品牌宣传册完美地体现了坐落在全球机场航空站内的喜力品牌以美食和饮品为理念的设计内容、设计目的和应用。该手册旨在引领读者自由出入该酒吧间的模块，清晰地勾勒出每一个元素，并为承租人和运营商提供完全渲染三维布局，以打造身临其境的视觉效果。
　　手册的设计和版式同样也遵循了品牌理念的外观和感觉。封面经紫外线高光处理，以亚光黑色原浆纸为原料，并巧妙地引入了酒吧间内部定制的纯黑色壁纸。

# Dasmusic Programation 2010-2011

"Dasmusic 音乐机构2010–2011年度规划"宣传册

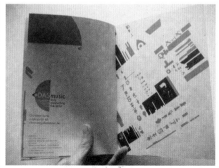

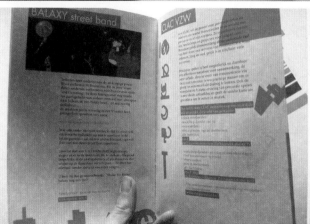

Design Agency:
**Jensdawn**
Creative Director:
**Jensdawn**
Designer:
**Jensdawn**
Photography:
**Sofie Jaspers**
Client:
**Dasmusic**
Nationality:
**Belgium**

设计机构：
真斯多恩设计工作室
创意总监：
真斯多恩
设计师：
真斯多恩
摄影：
索菲亚·雅斯贝尔斯
客户：
Dasmusic 音乐机构
国家：
比利时

Dasmusic is a music agency that asked Jensdawn to review the logo and make a booklet with information about the musicbands behind Dasmusic. The colours of the original logo were blue and black, so the designer stayed with these colours throughout the book. He tried to make an abstract language with objects, like notes on a musicsheet. The inside and outside of the cover fit together as one big image, and that big image shows all the pages, puzzled and scalled down(the texts are the black lines).

"Dasmusic"是一个音乐机构，委托真斯多恩设计工作室为其标识进行重新设计，并制作一个宣传手册以介绍该机构背后的音乐团队。原有标识的色调以蓝色和黑色为主，因此，设计师在书中保留了这些色调。设计师尝试运用物体创建一个抽象的语言，犹如乐谱上跳动的音符。
封面内外完美结合后形成一个大幅图案，而这一图案出现在所有的页面之中，犹如迷宫一般发人深思（黑线部分是文字）。

# Ruff&Cut Brochure

Ruff&Cut公司宣传册

Design Agency: **Imaginaria Creative**
Creative Director: **Cesar Sanchez**
Designer: **Cesar Sanchez**
Client: **Ruff&Cut**
Nationality: **USA**

设计机构：Imaginaria创意设计工作室
创意总监：塞萨尔·桑切斯
设计师：塞萨尔·桑切斯
客户：Ruff&Cut公司
国家：美国

Using only conflict - free diamonds, which are untampered with by human hands, Ruff&Cut is an eco-friendly, socially responsible and ethical jewellery company that creates beautiful jewellery with recycled metals in organic shapes. This brochure helps to capture the raw essence of the landscape and the beauty of the jewellery itself.

Ruff&Cut公司是一个注重环保和社会责任感的珠宝公司，强调手工设计的自由雕刻、无添加任何修饰，以有机形态的回收金属为原料打造出美轮美奂的珠宝。该手册的设计完美体现了该公司的纯粹理念以及珠宝的天然之美。

# Cautiva

"俘虏"宣传册

Design Agency:
17-30.com
Creative Director:
Javier Latorre / Lucía Meseguer
Designer:
Javier Latorre / Lucía Meseguer
Client:
Innova
Photography:
Bruno Almela / Miguel Barguês / Inma Riquelme / Fernando Fernández / Paco Latorre
Nationality:
Spain

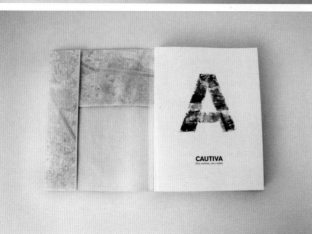

设计机构:
17-30.com设计工作室
创意总监:
哈维尔·拉特来,露西娅·梅塞格
设计师:
哈维尔·拉特来,露西娅·梅塞格
客户:
英诺华公司
摄影:
布鲁诺·爱尔梅拉,米盖尔·巴格斯,伊玛·里克尔梅,费尔南多·费尔南德斯,帕科·拉特来
国家:
西班牙

Captive, the project is a reaction, a personal expression to a problem that hits the critical attitude of many in Valencia, not all. Captive is a bipolar journey through the city of Valencia; the book guides a tour of the city "more real" and less invented, the city with which the designers identify the authors, there is, that no makeup or opens, which account for treasure their neighbourhoods, their estate buildings and its people for fuel.
The development of the publication shows this view of the city to finish the ride with a longing for what is and what he wants to be, what remains to be done and the direction the designers want to do. In the book an invaluable collaboration is made with pictures and articles: Bruno Almela, Miguel Bargues, Inma Riquelme, Fernando Fernández and Paco Latorre.

该项目是对社会的一个反思，彰显了个人对瓦伦西亚某些（非全部）决定性态度的见解。该项目是一个有关瓦伦西亚的"极端"旅行，它将引领读者感受一个"更为真实"的城市。设计师旨在彰显一个没有伪装的城市或个人，强调对周围环境、房产建筑以及民众的关注。

该出版物的设计展现了城市的属性、发展前景以及存在的问题和设计师对城市发展的观点。在书中，设计师巧妙运用了大量图片和文章，这些内容由布鲁诺·爱尔梅拉，米盖尔·巴格斯，伊玛·里克尔梅，费尔南多·费尔南德斯，帕科·拉特来提供。

# EPOC Handcrafted Beds Brochure

EPOC手工制作床品宣传册

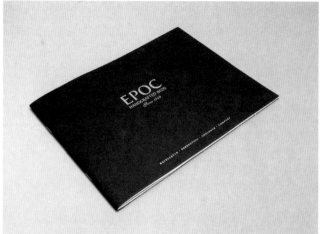

This is a high quality brochure to showcase EPOC Handcrafted Beds product range with a gold foil blocked cover on thick stock and lavish photography throughout. The brochure has a luxurious feel to complement the products - bespoke handmade beds for the most discerning customer.

这是一个设计精美的产品手册,旨在展示EPOC手工制作床品公司的产品线。烫金封面以厚实的原浆纸为原料,丰富的图片贯穿整个手册。手册的奢华之感与产品的风格相得益彰,即使是最挑剔的顾客也能够在这里挑选到称心如意的商品。

Design Agency:
Circleline Design Ltd.
Creative Director:
Martin Daines
Designer:
Owen Mathers
Client:
Mansion House Bedding Company
Photography:
Supplied
Nationality:
UK

设计机构:
Circleline设计有限公司
创意总监:
马丁·戴内斯
设计师:
欧文·玛德斯
客户:
广厦床品公司
摄影:
广厦床品公司
国家:
英国

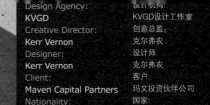

Design Agency: 设计机构:
**KVGD** KVGD设计工作室
Creative Director: 创意总监:
**Kerr Vernon** 克尔弗农
Designer: 设计师:
**Kerr Vernon** 克尔弗农
Client: 客户:
**Maven Capital Partners** 玛文投资伙伴公司
Nationality: 国家:
**UK** 英国

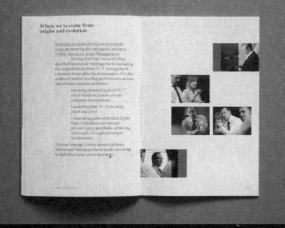

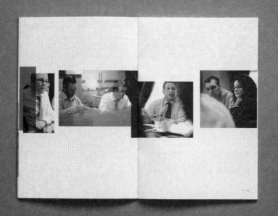

The brand launch involved logo design, stationery, a 20-page document and content managed website. Into the mix went classic typography, black and white photography, and highend print production values such as foil blocking and letterpress printing. The result is a brand and style that's contemporary and timeless.

这一品牌的开发方案涉及标识、文具用品、20页文献以及托管网站的设计。设计师巧妙地将经典的字体和黑白色调的照片以及诸如烫金和字母压花等高端印刷制作理念完美融合。最终的设计结构是巧妙地打造了一个时尚、永恒的品牌和风格。

# Francedanse Festival Brochure  Francedanse舞蹈节宣传册

Designer:
**Alexander Kalachev**
Client:
**French Institute in St. Petersburg**
Photography:
**Dance Companies**
Nationality:
**Russia**

设计师：
亚历山大·卡拉切夫
客户：
圣彼得堡法国学院
摄影：
舞蹈公司
国家：
俄罗斯

Francedanse – contemporary dance festival hosted within France- Russia year cultural programme. 7 performances including 2 special acts with French choreographers and Russian dancers plus outstanding play of famous choreographer Angelin Preljocaj inMariinsky Theatre took place within festival programme.
Brochure was created in one visual solution with all advertising materials, but by different design for each spread the designers wanted to show unique mood of every play.

"Francedanse"是在法国-俄罗斯年度文化节目中举办的当代舞蹈节。7个表演节目包括法国舞蹈编导和俄罗斯舞蹈家的两个特色演出，著名舞蹈编导Angelin Preljo c aj编排的普在马林斯基剧院卜演的杰作等。
该手册运用所有的广告宣传材料，构建出一个单一的视觉方案,但同时，设计师在每个跨页中运用的不同的设计手法，确保每个演出能够匠心独运。

# Green + Gold Printing Promo Brochure

绿色+金色印刷厂宣传手册

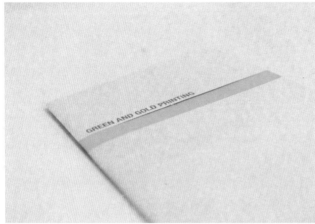

Design Agency:
Team Scope
Creative Director:
Mark Burrough
Designer:
Mark Burrough
Client:
Green + Gold Printing
Photography:
Lee Valentine
Nationality:
Australia

设计机构:
团队视野设计工作室
创意总监:
马克·博拉夫
设计师:
马克·博拉夫
客户:
绿色+金色印刷厂
摄影:
李·瓦伦蒂安
国家:
澳大利亚

This brochure for Green and Gold printing represented a great opportunity to really showcase their expertise. A combination of tricky printing techniques and an impressive portfolio of their finished work gave Green and Gold a great chance to let their "True Colours" shine through.

这个专为绿色+金色印刷厂而设计的宣传手册完美展示了该公司的专业资质。巧妙的印刷技巧和令人印象深刻的项目组合为该公司创造了一个伟大的展示机会,可以说,该手册是彰显公司"闪光点"的关键性元素。

# MacDonald-Miller Rebranding Programme

"麦克唐纳·米勒品牌重塑方案"宣传册

Design Director:
**Ron Lars Hansen**
Designer:
**Ron Lars Hansen**
Client:
**MacDonald-Miller**
Nationality:
**USA**

设计总监：
罗恩拉丝·汉森
设计师：
罗恩拉丝·汉森
客户：
麦克唐纳·米勒公司
国家：
美国

A Northwest leader in environmental systems, MacDonald-Miller engaged Hansen Belyea to reposition the company's brand to emphasize innovative engineering and operational excellence. The scope of work was extensive, including a robust new website, print collateral, advertisements, trade show graphics and specialty items. Without touching MacDonald-Miller's iconic red oval logo, Hansen Belyea retooled the company's messaging and brand identity system. Today the company is presented as a key player in forward-thinking engineering and energy efficiency programmes.

麦克唐纳·米勒公司是美国西北部环境系统的领军人物，他们委托汉森·贝尔雅设计工作室为其公司的品牌进行重新定位，以突出该公司的创新形式管理以及卓越经营模式。该项目的设计范围广泛，涉及全新网站、印刷材料、广告、展销会平面和专业物品的设计等等。设计师在保留该公司红色椭圆形标志的同时，对公司的信息传送和品牌识别系统进行了重组。如今，该公司在前瞻式管理和能源效率规划中扮演了一个重要的角色。

# 10 Years erinnern.at

"erinnern.at网络平台十年回顾"宣传册

Design Agency:
Sägenvier DesignKommunikation
Creative Director:
Sigi Ramoser / Cornelia Wolf
Designer:
Cornelia Wolf
Client:
erinnern.at, Bregenz
Nationality:
Austria

设计机构:
Sagenvier 视觉传达设计工作室
创意总监:
喜姬·拉莫赛,
科尼利亚·沃尔夫
设计师:
科尼利亚·沃尔夫
客户:
布雷根茨erinnern.at公司
国家:
奥地利

The Internet platform www.erinnern.at provides teaching and learning resources about the Holocaust and National Socialism, organises teacher education courses, and brings teachers together with researchers and institutions. The brochure is a documentation of their 10-year works.

因特网平台www.erinnern.at提供有关"大屠杀"和"国家社会主义"的教学和学习资料,组织师资培育课程,并建立教师与研究人员和机构之间的联系。该手册是一部记录该平台10年间工作情况的历史文献。

# Fdação Gonçalo da Silveira Institutional Brochure 葡萄牙耶稣会学会宣传册

Design Agency:
**Loja das Maquetas**
Designer:
**Cátia Alves**
Client:
**Fundação Gonçalo da Silveira**
Nationality:
**Portugal**

设计机构：
Loja das Maquetas设计工作室
设计师：
凯茜·阿尔维斯
客户：
葡萄牙耶稣会学会
国家：
葡萄牙

Taking the hand symbol as the most important reference for FGS, a new pitch line was conceptualise to promote the foundation work: "Lend a hand to those in need". All communication pieces had to connect to this new idea, a layout that would carry the principle of one giving a hand to someone in need.

该手册以手掌标志作为葡萄牙耶稣会学会的典型形象，概念化的节线有效地宣传了该学会的主旨，即"对有需要的人伸出援助之手"。所有的传达设计方案均与这一全新理念完美衔接，独特的版式设计将"对有需要的人伸出援助之手"的理念贯彻得恰到好处。

# SMO Brochure

SMO公司宣传册

Design Agency:
Sägenvier DesignKommunikation
Creative Director:
Sigi Ramoser / Cornelia Wolf
Designer:
Cornelia Wolf
Client:
SMO, Bregenz
Nationality:
Austria

设计机构：
Sagenvier 视觉传达设计工作室
创意总监：
喜姬·拉莫赛，科尼利亚·沃尔夫
设计师：
科尼利亚·沃尔夫
客户：
布雷根茨SMO公司
国家：
奥地利

The SMO companies offer neurological rehabilitation. Their patients suffer from the aftermath of a stroke, multiple sclerosis or Parkinson. The brochure explains their therapies and services.

SMO公司以提供神经康复治疗为主旨。来到该公司的患者往往是遭受中风后遗症、多发性硬化症或帕金森氏症等疾病的困扰。该宣传册主要向读者诠释他们的治疗方法和服务项目。

# Palíndromo #1 "1号回文诗"宣传手册

Design Agency:
**Greco Design**
Creative Director:
**Gustavo Greco**
Designer:
**Tidé, Fred Fita, Ana Luiza Gomes, Zumberto, Ricardo Donato, Fernanda Monte-Mor, Daniele Pires, Bruna Ladeira, Alexandre Fonseca**
Client:
**Rona Editora**
Nationality:
**Brasil**

设计机构：
Greco设计公司
创意总监：
古斯塔沃·格雷科
设计师：
蒂德、弗莱德·菲塔、阿纳·路易莎、戈梅斯、赞伯托、里卡多·多纳托、费尔南达·蒙特–莫尔、达尼埃尔·皮雷、布鲁纳·拉迪拉、亚历山大·丰塞卡
客户：
罗纳艾迪托拉公司
国家：
巴西

The purpose of Palíndromo is to promote the portfolio of Rona Editora, a traditional company in the graphics industry in Minas Gerais. A flexible format was created for the publication and interdisciplinary collaboration was proposed for the production of the content of each edition. Distinct types of paper and finishes were chosen in order to present the diversity of graphics services and the richness of the discussion. Palíndromo, the name given to the project, comes from Greek and means repetition or return. Each year a new trail, path, a new way of publishing words and images, another method of "writing" and "reading", of recording and learning is launched.

"回文诗"宣传手册的设计旨在对富有悠久历史的图像公司——罗纳艾迪托拉公司进行宣传。设计采用了灵活的版式，在各个不同的章节设计中融入了跨学科合作。独特的纸张和装饰呈现了公司图像服务的多样性。"回文诗"一词来自于希腊语，意为"重复、循环"，象征着公司每年都会开发新的道路、新的图文出版方式和新的写作阅读方法。宣传手册将这些前进的足迹记录了下来。

# Easy Guide - Hotel Wiesler

威斯勒酒店指南手册

Design Agency:
**moodley brand identity**
Creative Director:
**Mike Fuisz**
Designer:
**Josef Heigl**
Client:
**Hotel Wiesler**
Photography:
**Lupi Spuma / Marion Luttenberger**
Nationality:
**Austria**

设计机构:
Moodley品牌识别设计工作室
创意总监:
麦克·弗赛兹
设计师:
约瑟夫·黑格尔
客户:
威斯勒酒店
摄影:
卢比·斯帕玛，马里恩·卢登博格
国家:
奥地利

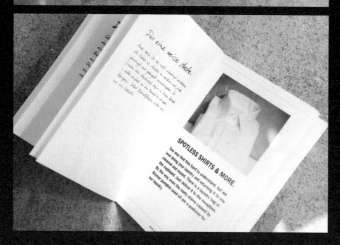

After a soft relaunch of the famous Austrian Hotel Wiesler, their guests have lately been pampered by thoughtful details totally free of formal pomp - and this is what the small hotel brochure, designed by moodley brand identity, is all about: containing opening hours and special offers in the in-house restaurants and bars, as well as the use of the telephone, air conditioning, Wi-Fi, laundry, how to organise wake up calls, etc. - all useful and practical matters are being decribed in the "Easy Guide" in a charming and unconventinal way. The pictures included have been taken in and around the hotel, and the handwriting in German is clearly set apart from the information in English.

著名的奥地利威斯勒酒店在重新营业之后，凭借周到、细致的服务使客人轻松摆脱了奢华空间所带来的拘谨之感。该酒店委托Moodley品牌识别设计工作室为其设计一款小型酒店手册，其中包括酒店营业时间、餐厅和酒吧内部的特价优惠、电话的使用说明、空调、无线上网、洗衣店以及叫醒服务等。该手册以一个迷人而非传统的方式将所有有帮助和实用的事项一一讲解。手册中所运用的图片均以酒店为主题，而德文的手写体则与英语信息内容鲜明地区分开来。

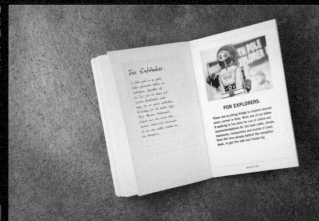

# Barbal Bascule Bridge Brochure Barbal公司"吊桥"宣传册

Barbal develops weighing technology namely bascules, scales and measuring systems. The visual language was developped to enlighten the technological and industrial character of the products and applied in the bascule bridge brochure with a clean layout, an elegant format and rich production details.

Barbal公司以开发称重技术,即吊桥活动桁架、天平和测量系统为主要特色。该视觉语言的设计旨在突出该公司产品的技术和工业特点,并与清晰的版式、优雅的开本和丰富的产品细节运用到吊桥宣传册之中。

| | |
|---|---|
| Design Agency: | 设计机构: |
| **Gen Design Studio** | Gen设计工作室 |
| Creative Director: | 创意总监: |
| **Leandro Veloso** | 林德罗·维罗索 |
| Designer: | 设计师: |
| **Carla Ribeiro** | 卡拉·里贝罗 |
| Client: | 客户: |
| **Barbal** | Barbal公司 |
| Photography: | 摄影: |
| **Leandro Veloso** | 林德罗·维罗索 |
| Nationality: | 国家: |
| **Portugal** | 葡萄牙 |

# Rinspeed Brochure 林斯比得公司宣传册

Creative Director:
**Caroline Sauter**
Designer:
**Caroline Sauter**
Client:
**Rinspeed Company**
Photography:
**Rinspeed Company**
Nationality:
**Germany / Swiss**

创意总监:
卡罗琳·索特
设计师:
卡罗琳·索特
客户:
林斯比得公司
摄影:
林斯比得公司
国家:
德国，瑞士

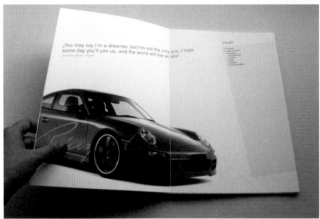

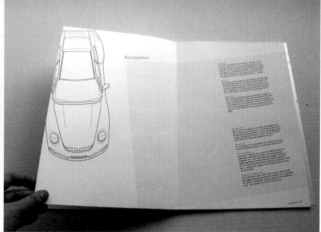
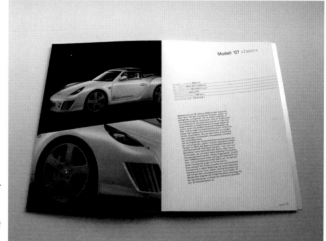

Rinspeed is a company for car tuning, specialist for porsche. Broshure is designed for their bussiness.

林斯比得公司是保时捷等汽车的专业改装中心。该宣传册专为林斯比得公司而设计。

# Metronet Sales Brochure   Metronet公司销售宣传册

Design Agency:
**Sensus Design Factory Zagreb**
Creative Director:
**Nedjeljko Spoljar**
Designer:
**Nedjeljko Spoljar / Kristina Spoljar**
Client:
**Metronet Telecommunications**
Photography:
**Photodisc**
Nationality:
**Croatia**

设计机构:
萨格勒布Sensus设计工厂
创意总监:
耐蒂艾尔克·斯颇里扎
设计师:
耐蒂艾尔克·斯颇里扎，克莉斯汀娜·斯颇里扎
客户:
Metronet电信公司
摄影:
Photodisc公司
国家:
克罗地亚

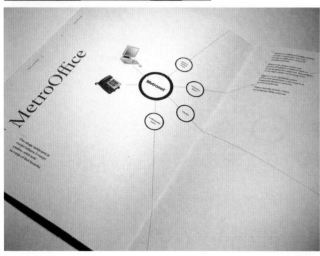

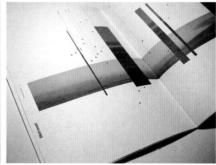

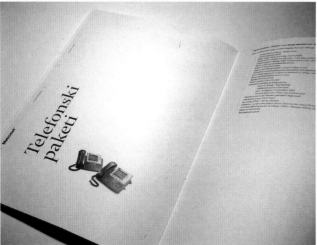

Sales brochure for a telecommunications company. Lavishly produced, using a paper with special tactile quality and feel, the brochure is aimed to attract big clients. Panoramic photographs and graphic "impulses" of irregular duration suggest the characteristics of firm's business, while illustrated diagrams allow for easier comprehension of the services. The position of dots appearing on the covers and inside corresponds to the distribution of the Metronet network on a map of Croatia.

该项目是专为Metronet电信公司而设计的销售宣传册。大量发行的该手册运用一个特殊质感的纸张为原料，旨在吸引大客户的注意力。全景相片和不规则持续的"脉冲"图案暗示了该公司业务的特性，而说明性图表则能够帮助读者对服务项目进行深入的了解。封面和内页中的圆点定位与Metronet公司在克罗地亚版图上分布的网络相得益彰。

# Publications for Mac/val Contemporary Art Museum
麦克/瓦尔当代艺术博物馆宣传册

**Design Agency:**
BURO-GDS with Zaijia Huang
**Designer:**
Ellen Tongzhou Zhao / Zaijia Huang
**Client:**
Mac/val Contemporary Art Museum
**Nationality:**
USA

设计机构：
BURO-GDS 设计工作室，黄哉佳
设计师：
艾伦·赵，黄哉佳
客户：
麦克/瓦尔当代艺术博物馆
国家：
美国

The design work includes documents, flyers, brochures, and labels for permanent and special exhibitions and events at the Mac/val Contemporary Art Museum.

该项目是专为麦克/瓦尔当代艺术博物馆中举办的永久性、特别展览和活动而设计的文献、传单、手册和标签。

# STRIP – Under the Skin "卸下束缚-肌肤下的秘密"宣传册

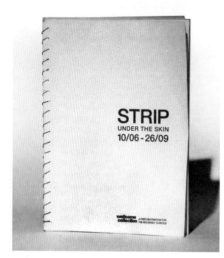

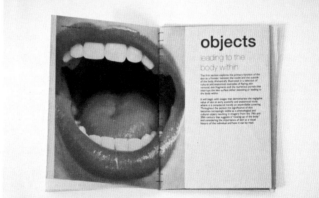

Creative Director:
Peter Su
Designer:
Peter Su
Client:
Wellcome Collection
Photography:
Peter Su
Nationality:
UK

创意总监:
彼得·苏
设计师:
彼得·苏
客户:
"维尔康姆的收藏"博物馆
摄影:
彼得·苏
国家:
英国

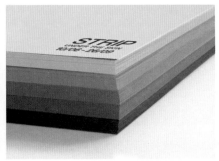
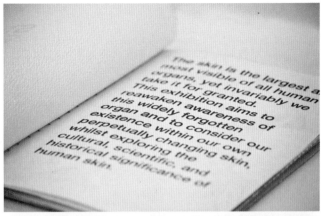

The project includes promotional materials for a forthcoming exhibition at Wellcome Collection. The theme of the exhibition is based on "Human Skin". The reason behind the name is because it can be defined as a process of removing something, playing on the idea of discovering more as you peel back the layers.

该项目是专为在"维尔康姆的收藏"博物馆即将举办的展览而设计的宣传材料。
该展览的主题围绕"人类皮肤"而展开。以此为主题的原因在于其可以被定义为移除障碍的过程，寓意深入内层发现更多不为人知的讯息。

# Almanaque do Samba-Jazz "桑巴-爵士音乐会年鉴"宣传册

Design Agency:
**Pós Imagem Design**
Creative Director:
**Rafael Ayres**
Designer:
**Daniel Escudeiro**
Client:
**Centro Cultural Banco do Brasil**
Nationality:
**Brazil**

设计机构：
Pos Imagem设计工作室
创意总监：
拉斐尔·艾尔斯
设计师：
丹尼尔·埃斯库德罗
客户：
巴西银行文化中心
国家：
巴西

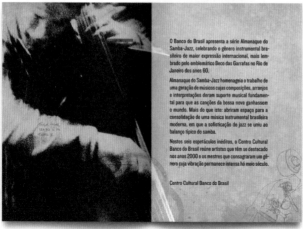

It is the brochure design for a series of Samba-Jazz concerts in Rio de Janeiro, Brazil.

该项目是专为在巴西里约热内卢举办的一系列桑巴－爵士音乐会而设计的宣传册。

## 16.MAR

# DAVID FELDMAN
## convida PAULO MOURA

Paulo Moura marcou com o microfone sua importante presença na discografia do samba-jazz, em grandes registros como Gismonti's Bossa Nova, O Trio 3-D Convida, Embalo e em seu Paulo Moura Quarteto. No Almanaque do Samba-Jazz de retomam este repertório e a sonoridade de seu clarinete e o instrumento que passou a adotar há alguns anos.

A carreira do jovem David Feldman segue a trilha dos músicos que se homenageia em seu disco de estréia, O Som do Beco das Garrafas, lançado em 2009. Estudioso do piano de nossos, formou-se em música no exterior, trabalhou este nomes do jazz internacional. De volta ao Brasil, tem atuado como arranjador, e mostra também seu próprio trio com o qual nos mostra um estilo original e se mesmo tempo reverente à força musical alinhada por seus mestres.

## 23.MAR

# SAMBAJAZZ TRIO
## convida MAURICIO EINHORN

Na audição de trios como Bossa Três, Tamba, Zimbo, Tempo, Sambrasa, Sambalanço e equilibrada triangulação entre piano, baixo e bateria fez a música brasileira dos anos 60 um caminho ao mais recente, bem atual e alternativo ao bamparandeira-viola.

Quarenta anos depois, o Sambajazz Trio reagrupa esta clássica formação instrumental. Seu disco Agora Sim é um dos mais expressivos registros da continuidade do samba-jazz nos anos 2000. Ao vivo, acrescenta à clássica formação o trompete tocado pelo bairista Clauton Sales. O resultado e fãs surpreenderão na comunidade pela performance, a mostra que com um trio de samba-jazz ora sempre pode ser mais.

Este espetáculo tem a participação especial de um mestre da guitarra na música brasileira. Mauricio Einhorn apresenta no Almanaque do Samba-Jazz temas de seus discos mais recentes, Transversos, de 2007. Mostra outras gravações no álbum ME, raro compilar de samba-jazz no final dos anos 70. E voltará a algumas de suas composições registradas nos anos 60 por colegas do disco clássico como No nossa Estrutura, do pianista Luiz Carlos Vinhas.

## 02.MAR

# HAMLETO STAMATO
## convida RAUL DE SOUZA

O pianista Hamleto Stamato viajou o mundo acompanhando grandes cantores e cantoras. Ao voltar o repertório do samba-jazz, os discretos como arranjador e intérprete de 2005 a 2008, graças Speed Samba Jazz, uma trilogia do álbum em que nos dois discos versões clássicas do gênero, e apresenta composições próprias como "Tema de Academia" e "Orlà", que escreveram ao lado da "Via Bargo", de autoria de Paulinho Trompete.

Raul de Souza retumante a maneira de tocar o trombone, ao mudar a afinação e colorir sua vibrata e mais seu instrumento. Esteve presente em todas as edições brasileiras do samba-jazz, como em Embalo, do pianista Tenório Jr., e Edison Machado É Samba Novo, do genial bateirista de mesmo nome. Para apostar ao vivo acompanha o Stamato seleccionaram seus gravados em À Vontade Mesmo, seu assunto mais marcante, e Bossa Eterna, seu disco mais recente.

## 09.MAR

# HENRIQUE BAND
## convida ANTONIO ADOLFO

Em 1965, os jovens e já conhecidos Paulo Moura, Meirelles, Rosolino e Edson Maciel participaram da gravação de O Trio 3-D Convida, credenciando o talento de um menino que afinava a música em três 3-D no ano anterior, quando – ainda menor de idade – se apresentou no "Little Club" e gravou seu primeiro disco, Tema 3-D. Desde então, o espaço reservado por diversas caminhos da MPB, tornando ainda mais especial sua passagem pelo samba-jazz, que será revivida neste Almanaque.

Ainda no final dos anos 60, Henrique Band foi criador do Onze Cabeças, pequena grande orquestra cujo nome já antecipou sua preferência do jovem arranjador e arranjador por grupos numerosos. Entre 2007 e 2009, gravou seu primeiro disco solo, Caleidoscópio, apresentando pelo samba-jazz e inserindo outras temas e orquestrações de música popular – uma busca acentuada por criar linhas brasileiras cuja referência musical conhecida e essencial Coisas, de Moacir Santos.

A apresentação de hoje da Henrique Band celebra o trabalho do arranjador com os maestros Cipó e Walfrido Bianchi, os Uns Uns, duos da bateria Dom Um Romão, e do maestro Walter em Marcos Tavares e Sarolta, quando seus arranjos de arranjador de canções de Hemi Manciel toda a sua orientação da música popular de samba-jazz.

O repertório revisitará encontros dos principais momentos da samba-jazz nos discos gravados nos anos 60, como no Sobolabrasas, disco do Em son Chaves, que tinha Raul de de Souza e Hector Costita entre seus dez craques, o LP de Os Cobras, experiente quarteto e quinteto o pianista Tenório Jr., o bateirista Milton Banana, o contrabaixista Zezinho Alves, e uma fornalha de sopros formado por Raulzinho, Meirelles, Paulo Moura e o trompete Hamilton Cruz.

## 30.MAR

# ALMANAQUE SAMBA-JAZZ BAND
## convida HECTOR COSTITA

Nosso convidado, o argentino Hector Bisignani, o Costita, é um grande músico brasileiro. Residente em nosso país há mais de cinquenta anos, gravou Inquieto, de composições próprias como o do título "Tangentes", em que o assunto-jazz visita a Argentina, atuou, executando estas e outras gravações em discos como No bandeada, de Erlon Chaves, Você Ainda Não Ouviu Nada, de Sergio Mendes e outros de Manfredo Fest, Luiz Chaves e o Sambalanço Trio aos quais Costita emprestou seus sopros.

A Almanaque Samba-Jazz Band foi criada especialmente para este apresentação: juiz repertório novo em disco ideia os maestros de Victor Assis Brasil e Canto de Edson Costita e seu poeta do Luiz Eça, a bossa do contrabaixo como Manoel Gusella, Tito Veto e Zezinho Alves, e os trios de Milton Banana e a piadação de Edson Machado celebrando o boneco que destaca na música atual.

Recebido maioris entre os mais importantes de seu projeto, em quarenta anos e alguma coisa, O Tabulo, Conjunto Cantoleiba, Quam Centeto, Os Polígonos, Os Cinco Peixes, Os Dois do Porto, que formam parte mais significativa presença na história do samba-jazz.

A homenagem aos artistas participantes, aos grupos brasileiros, aos autores do repertório apresentado, a seus arranjadores e instrumentistas convocados, entre Almanaque do Samba-Jazz se expande por esse historia nunca contada.

# Swedish Birth Day Brochure

瑞典"诞生之日"宣传册

Design Agency:
**Turnstyle**
Creative Director:
**Ben Graham**
Designer:
**Madeleine Eiche**
Client:
**Swedish Hospital**
Photographer:
**Bella Baby**
Nationality:
**USA**

设计机构:
Turnstyle设计工作室
创意总监:
本·格雷厄姆
设计师:
玛德琳·爱其
客户:
瑞典医院
摄影:
贝拉婴儿公司
国家:
美国

As a parent-to-be, one of the first important decisions you'll make is where to have your baby. This brochure, targeted at expectant parents, serves as an introduction to the Swedish Women and Infant Centre's comprehensive range of services.

对于准爸妈们来说,其中一个最重要的决定就是选择孩子的诞生地。该手册以准爸妈们为对象,向读者介绍了瑞典妇婴中心的综合性服务。

# Design Week   "设计周"宣传册

Design Agency:
**Rethink**
Creative Director:
**Ian Grais / Chris Staples / Jeff Harrison**
Designer:
**Ian Grais / Jeff Harrison / Todd Takahashi / Rory O'Sullivan / Leia Rogers / Matt Warburton / Steve Fisher / Jay Grandin / Mark Busse / Rod Roodenberg / Paul Bazay**
Nationality:
**Canada**

设计机构：
新理念设计工作室
创意总监：
伊恩·格雷斯，克里斯·斯台普斯，杰夫·哈里森
设计师：
伊恩·格雷斯，杰夫·哈里森，托德·高桥，洛伊·奥沙利文，莱亚·罗杰斯，马特·沃伯顿，史蒂夫·费舍尔，杰·格兰丁，马克·布塞，罗德·卢登堡，保罗·贝基
国家：
加拿大

Large questions throughout the Design Week 2010 guide were used to help attendees start conversations with each other about just how you go about defining the value of design.

　　2010年设计周指南的设计旨在帮助参加者之间进行有效的沟通，陈述个人对设计价值的理解。

# Brochure for Energoholding   Energo-Holding集团公司宣传册

Creative Director:
**Roman Krikheli**
Designer:
**Andrey Koodenko / Andrey Ilyaskin**
Client:
**Energo-Holding Group of Companies**
Photographer:
**Roman Krikheli**
Nationality:
**USA**

创意总监：
罗马·克里克赫立
设计师：
安德烈·肯德克，安德烈·伊雷亚斯金
客户：
Energo-Holding集团公司
摄影师：
罗马·克里克赫立
国家：
美国

The designers made this brochure for a Russian Energetics holding using NASA images, film photography and neat typography. They succeeded in producing a very functional and edgy design at the same time. The brochure is mainly used as a presentation material and it serves its purpose very well for managers. Designers like it too - taking into account the top-notch imagery and design.

该项目是专为一个俄罗斯动力学工程控股公司而设计的手册，该手册中大量运用了美国国家航空航天局的形象、摄影胶片以及整洁干练的字体。该设计兼具实用性和美观的特点。这一手册主要作为一种有效的宣传材料，为管理者提供良好的辅助作用。此外，精美的图案和设计风格令设计师自身爱不释手。

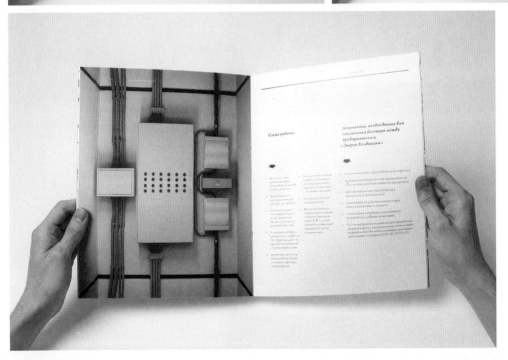

# Honda: Design for Russians Research

"本田:专为俄国人而设计的调查报告"宣传册

Design Agency:
**Communication Bureau Proekt**
Creative Director:
**Roman Krikheli / Alexey Bakirov**
Art Director:
**Alexey Novikov**
Client:
**Honda**
Nationality:
**Russia / Japan**

设计机构:
Proekt传达设计工作室
创意总监:
罗马·克里克赫立, 阿克列谢·巴克洛夫
艺术总监:
阿克列谢·巴克洛夫
客户:
本田公司
国家:
俄罗斯, 日本

Honda - being one of the largest automobile manufacturer—launched a multi-national tender process for a research to point out people's attitudes towards design. The cause of it is that Honda was planning to start working in premium segment of the market. In Russian contest the designers won thanks to a non-standard approach: they didn't ask just regular people, they made a few interviews with top Russian designers and trend-setters so that the information would be more accurate and to the point. The resulting brochure was a great success.

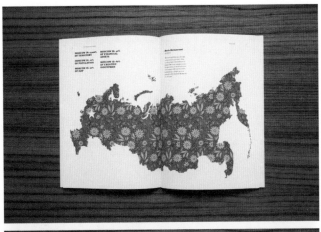

全球最大的汽车制造商本田公司发起了一项跨国招标方案，旨在对人们对设计所作出的反应而进行调查。该调查方案的设计主要是源自本田公司放眼高档汽车市场的规划。在俄罗斯团队大赛中，Proekt传达设计工作室凭借一个非标准的设计手法而获胜。该工作室对普通民众和少数的顶尖俄罗斯设计师和流行时尚引导者进行了细心的调查，从而获得了更为确切的信息，切中要点。该手册是一个非常成功的项目。

# The Overlake School Brochure  Overlake学校宣传册

Creative Director:
**Ron Lars Hansen**
Designer:
**Ron Lars Hansen**
Client:
**The Overlake School**
Nationality:
**USA**

创意总监：
罗恩拉丝·汉森
设计师：
罗恩拉丝·汉森
客户：
Overlake学校
国家：
美国

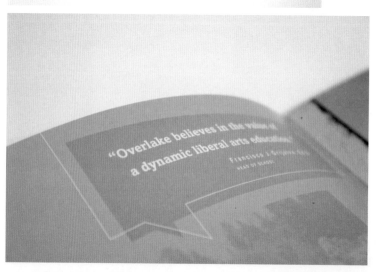

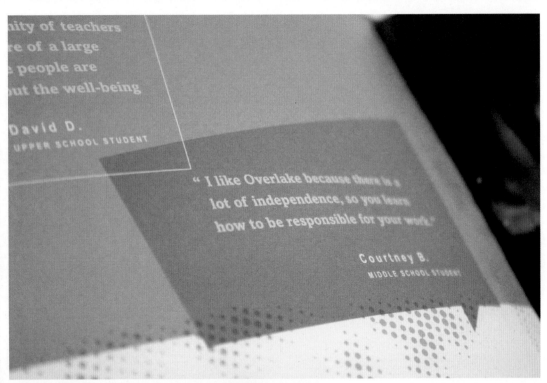

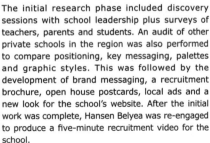
The initial research phase included discovery sessions with school leadership plus surveys of teachers, parents and students. An audit of other private schools in the region was also performed to compare positioning, key messaging, palettes and graphic styles. This was followed by the development of brand messaging, a recruitment brochure, open house postcards, local ads and a new look for the school's website. After the initial work was complete, Hansen Belyea was re-engaged to produce a five-minute recruitment video for the school.

最初的调研阶段包括与学校领导进行的研讨以及在教师、家长和学生之中的调查。此外，对该地区其他私立学校的调查也为该学校的定位、关键信息传送、配色方案以及平面风格奠定了设计基础。随后，设计师还为该学校开发了品牌信息传送、招聘手册、开放日明信片、局部广告以及更新的学校网站设计方案。在初期工作全部完成之后，汉森·贝尔雅设计工作室再次受委托为该学校设计一个五分钟的招聘视频。

# Magaggiari Hotel Resort - Brochure
Magaggiari度假酒店宣传册

Design Agency:
**OOCL - Creativity Addicted**
Creative Director:
**OOCL - Creativity Addicted**
Designer:
**Marco Lorio**
Copywriter:
**Francesco Colantonio**
Client:
**MHR - Magaggiari Hotel Resort**
Photography:
**Giorgio Vacirca**
Nationality:
**Italy**

设计机构：
OOCL – Creativity Addicted
创意设计工作室
创意总监：
OOCL – Creativity Addicted
创意设计工作室
设计师：
马可·罗里奥
文案：
弗朗西斯科·柯兰托尼奥
客户：
Magaggiari度假酒店
摄影：
乔治·瓦希尔卡
国家：
意大利

Elegance and simplicity are the essential features of Magaggiari Hotel Resort, the 4-star resort aiming to be the new pearl of the Sicilian coast near Palermo. Situated in a beautiful natural habitat among the mountains and the sea, MHR is a modern and comfortable relais ideal for peace and relaxation. The 16-page brochure, made with UV coating in total white, is part of the new corporate image of MHR.

　　高雅、简约是Magaggiari度假酒店的基本特色，这一四星级度假酒店一直努力成为巴勒莫附近的西西里海岸一颗璀璨的珍珠。该酒店坐落于一个美妙的自然环境之中，依山傍海，时尚、舒适的空间环境是客人寻找宁静和放松的理想港湾。这个16页的手册，以纯白色为基调，运用紫外线涂层材料，是该酒店全新企业形象的一个部分。

# Kallista Holding Profile　卡里斯泰控股公司宣传册

**Design Agency:**
Paragon Marketing Communications
**Creative Director:**
Konstantin Assenov
**Designer:**
Huzaifa Kakumama
**Client:**
Kallista Holding Co.
**Nationality:**
Bulgaria

设计机构：
典范市场营销设计工作室
创意总监：
康斯坦丁·埃塞诺夫
设计师：
胡扎艾法·卡库马玛
客户：
卡里斯泰控股公司
国家：
保加利亚

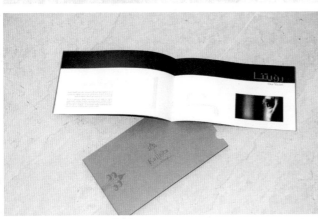

The profile uses the telegraph style of minimal approach although maintaining the high class and standard of the Holding Company, which deals in jewellery, watches, gift items and accessories. The profile is all about more in less approach. The entire profile with its jacket is 2 spot colours with UV.

该宣传册运用电报风格，以简约的手法传达出卡里斯泰控股公司产品的上乘品质和标准。该公司以经营珠宝、手表、礼品及配件为主要特色。该手册以体现简约的手法为主旨。整个作品与封套一同运用了紫外线材料，选用两种专色色调。

# Mataharis  "女私家侦探"宣传册

Design Agency:
**Barfutura**
Designer:
**Sergio Gonzalez Kuhn**
Client:
**Producciones La Iguana**
Nationality:
**Spain**

设计机构:
Barfutura设计工作室
设计师:
塞尔吉奥·冈萨雷斯库恩
客户:
Producciones La Iguana公司
国家:
西班牙

The design includes original film poster and pressbook for the Spanish movie "Mataharis", Official Selection at the San Sebastian International Film Festival 55th edition. Directed by Icíar Bollaín, the film explores relationships and ethics in the daily life of three working women.

该项目是专为第五十五届圣塞巴斯蒂安国际电影节上的参赛影片——西班牙电影《女私家侦探》而设计的原版影片海报和宣传资料。该电影由伊希娅·博拉茵编导,影片探讨了三个职业女性日常生活中的人际关系和伦理情感。

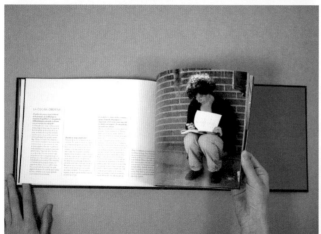

# Lol (Laughing Out Loud)  "LOL (哈哈大笑)"促销单

Design Agency:
**Barfutura**
Designer:
**Sergio Gonzalez Kuhn**
Client:
**Altafilms**
Nationality:
**Spain**

设计机构：
Barfutura设计工作室
设计师：
塞尔吉奥·冈萨雷斯库恩
客户：
Altafilms公司
国家：
西班牙

LOL (LAUGHING OUT LOUD) / Press book includes film pressbook and promotional pins / badges.

LOL (哈哈大笑)/ 宣传资料。电影宣传资料和促销徽章。

# The New Shoppingbag  新购物袋公司周年纪念宣传材料单

Design Agency:
**Studio Beige**
Client:
**Susan BijlWork**
Nationality:
The Netherlands

设计机构:
贝琪设计工作室
客户:
苏珊·贝吉尔设计工作室
国家:
荷兰

In 2008 The New Shopping bag celebrated its 5th birthday, and Studio Beige made a paper to celebrate that. It was very important to make the design eco-friendly, so the designers used eco-ink, FSC-certified and recycled paper. They folded the spreads together like a newspaper, to avoid using staples. A striking and simple graphic design that will hopefully get a great follow up! Award: Publication, prize winner EDAWARDS 2009.

2008年，为庆祝新购物袋公司成立五周年，贝琪设计工作室受邀为其提供了一个纸质宣传材料。为体现设计的环保意识，设计师巧妙运用了生态油墨、美国联邦科学委员会认证再生纸等材料。跨页经折叠后犹如一张报纸，避免了原料的浪费。醒目而简约的平面设计必定会获得受众的支持，从而引领一个全新的风尚。该项目曾荣获2009年EDAWARDS出版物金奖。

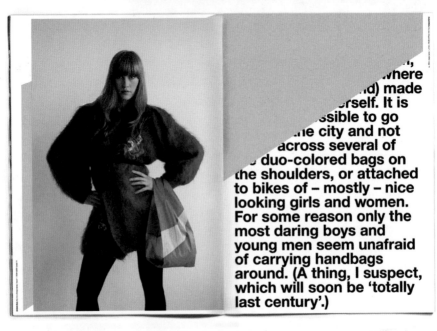

# Yum Cha and Me  "饮茶与我"宣传单

Design Agency:
**Toby Ng Design**
Creative Director:
**Toby Ng**
Designer:
**Toby Ng**
Nationality:
**Hongkong, China**

设计机构:
**托比·耐戈设计工作室**
创意总监:
**托比·耐戈**
设计师:
**托比·耐戈**
国家:
**中国香港**

It is talking about cultural identity. Yum Cha – literally means "drink tea", an activity core to the Cantonese culture, popped up to the designer. More than having dim sum and Chinese tea, Yam Cha is about family and union. One side of this concertina fold of card is entitled the designer's Dim Sum Family, where each of his family members are represented by a particular dim sum, which discloses their characteristics and relationship with him. The other side is called The Secret To Yum Cha, where some rituals of Yum Cha that may not be known to many are revealed.

该项目以文化识别为主题。"Yum Cha"从字面上可理解为"饮茶",强调中国的广东文化。与点心、中式茶相比,"饮茶"更强调家庭和团体性。该手册的一侧,详细介绍了设计师的茶饮家族成员,并详细介绍了每一样器具的特色和与设计师之间的联系。另一侧则被称为"饮茶中的秘密",其中介绍了很多不为人知的"饮茶"规矩。

# Korte Pocket Folder    科特公司口袋折页

Design Agency:
**Knoed Creative**
Client:
**The Korte Company**
Nationality:
**USA**

设计机构：
**科诺德创意设计工作室**
客户：
**科特公司**
国家：
**美国**

Korte needed a folder designed to fit a stack of successful case studies for potential clients. Because they're all about building smart, their pocket folder had to have the same integrity. Combining form, function and information, this 8-panel folder unfolds to show a timeline on one side and 2 deep pockets on the other.

科特公司需要设计一个折页以将他们的成功案例分析分发给潜在客户。由于该公司以建筑为特色，因此，该口袋折叠手册也同样围绕这一主题展开设计。这个8个模块折叠手册巧妙地将形态、功能和信息相结合，一端展示的是案例的年表，而另一端则是2个对案例的详细介绍的折页。

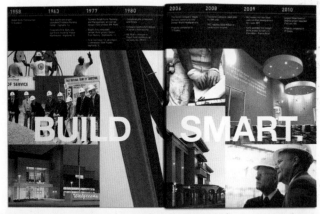

# Cilantro Café In-store Menu 希兰特罗咖啡馆店内菜单

Design Agency:
**Equinox Graphics**
Creative Director:
**Nermine Hammam**
Designer:
**Mohamed Eissa**
Client:
**Cilantro Cafe**
Photography:
**Mohamed Eissa**
Nationality:
**Egypt**

设计机构:
Equinox平面设计工作室
创意总监:
耐迈因·哈曼
设计师:
穆罕默德·埃塞
客户:
希兰特罗咖啡馆
摄影:
穆罕默德·埃塞
国家:
埃及

It is the menu design for Cilantro Café. This project uses the unique shape to reflect its special, the beautiful colours and lovely patterns make it more interesting.

该项目是专为希兰特罗咖啡馆而设计的菜单。该项目运用独特的形态以彰显其独特魅力,美妙的色调和可爱的图案令整个设计妙趣横生。

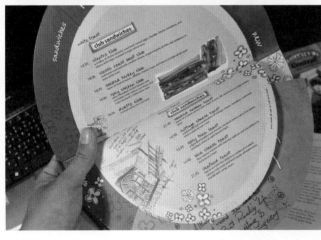

# KURO by Panamo

帕纳莫公司旗下酷乐餐厅的宣传单

Design Agency:
**ARTENTIKO**
Creative Director:
**Marcin Kaczmarek**
Designer:
**Marcin Kaczmarek**
Client:
**Panamo**
Photography:
**Marcin Kaczmarek**
Nationality:
**Poland**

设计机构:
ARTENTIKO品牌设计工作室
创意总监:
马克辛・凯克兹马莱克
设计师:
马克辛・凯克兹马莱克
客户:
帕纳莫公司
摄影师:
马克辛・凯克兹马莱克
国家:
波兰

For Japanese and Thai restaurant KURO by Panamo, ARTENTIKO made a professional logo facelift. The designers streamlined the form of the Japanese symbol of black colour (KURO) and selected a clearer and more minimalist typography. Then, as part of more extensive branding actions, the designers designed original promotional leaflets with menu.

对于由帕纳莫公司经营的日式与泰式主题餐厅－酷乐餐厅的品牌设计，来自ARTENTIKO品牌设计工作室的设计师为其精心打造了一个专业化的标识升级改造方案。设计师马克辛·凯克兹马莱克等人对黑色日本标志的形态进行了适当的简化处理，并选择了一个更为清晰、简约的版式格局。随后，为了更好地树立品牌形象，设计师们还专门为该餐厅量身打造了一个带有菜单的宣传单，设计风格独特、巧妙。

# 13º Festival Internacional de Curtas de Belo Horizonte 第十三届贝罗哈里桑塔国际短片电影节宣传单

Design Agency:
**Greco Design**
Creative Director:
**Gustavo Greco**
Designer:
**Tidé, Ana Luiza Gomes, Zumberto, Cláudio Silvano, Flávia Siqueira, Laura Scofield, Alexandre Fonseca**
Client:
**FundaçãoClóvis Salgado**
Nationality:
**Brasil**

设计机构:
Greco设计公司
创意总监:
古斯塔沃·格雷科
设计师:
蒂德、阿纳·路易莎·戈梅斯、赞伯托、克劳迪奥·希尔瓦诺、弗拉维亚·西科伊拉、劳拉·斯科菲尔德、亚历山大·丰塞卡
客户:
克洛维斯基金会
国家:
巴西

This project is the promotional material for the 13th Belo Horizonte International Short Film Festival, organised by FundaçãoClóvis Salgado at the Cine Humberto Mauro theatre. The works in the Belo Horizonte Short Film Festival were inspired by a great Brasilian filmmaker: Humberto Mauro. In order to give rhythm to the narrative of his films, the artist explored landscapes, always revealing the mountains and valleys so typical of Minas Gerais. A typographical family was created in order to lend support to the graphic pieces, using the proportions of a triangle, the graphic symbol of Minas Gerais.

该项目是为第十三届贝罗哈里桑塔国际短片电影节所设计的宣传单，电影节由克洛维斯基金会在温贝托·莫罗剧院举办。贝罗哈里桑塔短片电影节上的作品深受巴西著名电影制作人温贝托·莫罗的影响。他经常利用米纳斯吉拉斯地区特有的高山和峡谷等景色来增加影片的节奏感。为了搭配图像设计，设计师特别开发了一套字体。它们在字体中融入了代表着米纳斯吉拉斯地区的三角形图案。

# INDEX 索引

17-30.com

21pixels Studio

35 Communications (now Conran Design

Group)

804© Agentur fuer visuelle

Kommunikation

83 Designhouse

Alex Boland

Alexander Kalachev

Alistair Stephens Design

ARTENTIKO

Avigail Bahat

Barfutura

Bellamy Studio

BENBENWORLD

Billy Blue Student

Britta Siegmund

BURO-GDS with Zaijia Huang

Carlos Ribeiro

Caroline Sauter

CASA REX

Charlotte Aal / Karin ter Laak

Circleline Design Ltd.

Communication Bureau Proekt

COOEE - Leon Dijkstra

CreativeAffairs

"Danielle Blay / Kelly Satchell / Ally Carter

/ Isaac

Konczak / Elliott Denny"

Dimis design

Elke Broothaers

Equinox Graphics

| | |
|---|---|
| Erretres | Kerr Vernon Graphic Design |
| Etoile Rouge | key2design |
| Exposure | Knoed Creative |
| Forma | Kolektiv Studio |
| Freddy Thomas | Kollor |
| Frost design | Koniak Design |
| Gas Creative Group | KVGD |
| Gen Design Studio | Lisa Penedo |
| Goiaba | Lldesign |
| Greco Design | Loja das Maquetas |
| Hyperakt | Ludovic Roth Design Studio |
| Imaginaria Creative | Maggie Tsao |
| Interabang | Mapas |
| Jens Dawn | "Matt Van Ekeren / Molly Hein / Catherine |
| Jo Hoctor | Grothe |
| Jordan Puopolo Design | Bicknell / Christina Rimstad" |

# INDEX 索引

Mattias Sahlén – Graphic Design & Art Direction

"Michael Merzlikar / Martin Embacher / Stephanie Rappl / Vera Svoboda"

Michelle Liando

Mind Design

Monogum Creative

moodley brand identity

Nachtbrakers + Elke Broothaers

Nicole LaFave

Noblegraphics Creative Studio

OOCL - Creativity Addicted

Paprika

Paragon Marketing Communications

Peter Su

PLAZM

Pós Imagem Design

PRIM PRIM

Rain Visual Strategy + Design

Rethink

Roman Krikheli

Ron Lars Hansen

Roosje Klap

Sägenvier DesignKommunikation

Sarmishta Pantham

sebastianhaeusler.design

Sensus Design Factory Zagreb

Shenkar - College of Engineering & Design

Strichpunkt Design

Studio "ONY"

Studio 3

Studio Baklazan

Studio Beige

Studio Borzak

Superultraplus Design Studio

SZMER

Tamer Köşeli

Team Scope

Thorleifur Gunnar Gíslason

Toby Ng Design

TODA

Trampoline Design Pty Ltd.

Turnstyle

UXUS

Walsh

www.ewencom.com

Xavier Encinas Studio

YellowDog

Zoo Studio

图书在版编目（CIP）数据

商务印刷品设计 /（塞尔）伊利亚·德拉吉斯克，
（塞尔）伊戈尔·米拉诺维奇编；常文心译. -- 沈阳：辽
宁科学技术出版社，2014.11
　　ISBN 978-7-5381-8382-5

　　Ⅰ. ①商… Ⅱ. ①伊… ②伊… ③常… Ⅲ. ①商务－
印刷品－设计 Ⅳ. ①TS801.4

　　中国版本图书馆CIP数据核字（2013）第274617号

出版发行：辽宁科学技术出版社
　　　　　（地址：沈阳市和平区十一纬路29号 邮编：110003）
印 刷 者：利丰雅高印刷（深圳）有限公司
经 销 者：各地新华书店
幅面尺寸：170mm×220mm
印　　张：17
字　　数：50千字
印　　数：1～3000
出版时间：2014年 11 月第 1 版
印刷时间：2014年 11 月第 1 次印刷
责任编辑：陈慈良　王丽颖
封面设计：周　洁
版式设计：周　洁　关木子
责任校对：周　文
书　　号：ISBN 978-7-5381-8382-5
定　　价：88.00元

联系电话：024-23284360
邮购热线：024-23284502
E-mail：lnkjc@126.com
http://www.lnkj.com.cn
本书网址：www.lnkj.cn/uri.sh/8382